C.A. Davis Jr. • A.M.V. Monteiro

Advances in Geoinformatics

Clodoveu Augusto Davis Jr.
Antônio Miguel Vieira Monteiro
(Eds.)

Advances in Geoinformatics

VIII Brazilian Symposium on GeoInformatics, GEOINFO 2006, Campos do Jordão (SP), Brazil, November 19-22, 2006

With 134 Figures

Springer

EDITORS:

C. A. Davis Jr.
Informatics Institute,
Pontifical Catholic University of Minas Gerais
Rua Walter Ianni, 255
31980-000 – Belo Horizonte – MG
Brazil
e-mail: clodoveu@pucminas.br

A. M. V. Monteiro
Image Processing Division/Space and Society Initiative,
National Institute for Space Research - INPE
Avenida dos Astronautas,
1758 – Jardim da Granja
12227-010 – São José dos Campos – SP
Brazil
e-mail: miguel@dpi.inpe.br

ISBN 978-3-642-09247-3 e-ISBN 978-3-540-73414-7

Springer is a part of Springer Science+Business Media
springeronline.com

Preface

This volume contains selected papers presented at the VIII Brazilian Symposium on Geoinformatics, GeoInfo 2006, held in Campos do Jordão, Brazil, November 19-22, 2006. The GeoInfo conference series[1], inaugurated in 1999, reached its eighth edition in 2006. GeoInfo continues to consolidate itself as a the most important reference of quality research on geoinformatics and related fields in Brazil.

GeoInfo 2006 brought together researchers and participants from several Brazilian states, and from abroad. Among the authors of the accepted papers, 20 distinct Brazilian academic institutions and research centers are represented, in a clear demonstration of the expansion of research groups on geoinformatics throughout the country. The acceptance rate of GeoInfo conferences is increasingly competitive, which, in our opinion, is a clear sign of the vitality of our research community.

The conference included special keynote presentations by Christopher Jones and Martin Kulldorff, who followed GeoInfo's tradition of attracting some of the most prominent researchers in the world to productively interact with our community, thus generating all sorts of interesting exchanges and discussions. Keynote speakers in earlier GeoInfo editions include Max Egenhofer, Gary Hunter, Andrew Frank, Roger Bivand, Michael Worboys, Werner Kuhn, Stefano Spaccapietra, and Ralf Güting.

For the preparation of this edition, authors were encouraged to improve the conference papers, by incorporating suggestions and insights from the discussion sessions that followed oral presentations at the symposium.

We would like to thank all Program Committee members, listed below, and additional reviewers, whose work was essential to ensure the quality of every accepted paper. At least three specialists contributed with their review for each paper submitted to GeoInfo.

Special thanks are also in order to the many people that were involved in the organization and execution of the symposium, particularly INPE's invaluable support team, led by Terezinha Gomes dos Santos, with the participation of Daniela Seki, Hilcéa Santos Ferreira, Janete da Cunha, and Thanisse Silva Braga.

---

[1] http://www.geoinfo.info

Finally, we would like to thank GeoInfo's sponsors, identified at the conference's web site. The Brazilian Computer Society (Sociedade Brasileira de Computação, SBC), which supports the symposium, is the organization behind a number of successful Brazilian conferences on computing, most of which merit international recognition. The Brazilian National Institute of Space Research (Instituto Nacional de Pesquisas Espaciais, INPE) has provided much of the energy that has been required to bring together this research community in the past, and continues to perform this role not only through their numerous research initiatives, but by continually supporting the GeoInfo events and related activities, including the preparation of this book.

Belo Horizonte and São José dos Campos, Brazil

The editors,

Clodoveu Davis
Program Committee Chair

Antônio Miguel Vieira Monteiro
General Chair

GeoInfo Organization

General Chair
Antônio Miguel Vieira Monteiro, INPE, Brazil

Program Committee Chair
Clodoveu Davis, PUC Minas, Brazil

Program Committee
Ana Carolina Salgado, UFPE, Brazil
Andréa Iabrudi, UFMG, Brazil
Andréa Rodríguez, Universidad de Concepción, Chile
Andrew Frank, Technical University of Vienna, Austria
Antônio Miguel Vieira Monteiro, INPE, Brazil
Bernhard Mitschang, Universität Stuttgart, Germany
Christelle Vangenot, Swiss Federal Institute of Technology, Switzerland
Christopher Jones, Cardiff University, United Kingdom
Cirano Iochpe, UFRGS, Brazil
Cláudia Bauzer Medeiros, Unicamp, Brazil
Cláudio Baptista, UFCG, Brazil
Cláudio Esperança, COPPE/UFRJ, Brazil
Frederico Fonseca, Penn State, USA
Gilberto Câmara, INPE, Brazil
Guoray Cai, Penn State, USA
João Argemiro Paiva, Oracle Corporation, USA
Jorge Campos, UNIFACS, Brazil
José Alberto Quintanilha, USP, Brazil
José Luiz de Souza Pio, FUA, Brazil
Jugurta Lisboa Filho, UFV, Brazil
Karla Albuquerque de Vasconcelos Borges, Prodabel, Brazil
Kathleen Hornsby, University of Maine, USA
Lubia Vinhas, INPE, Brazil
Marcelo Tílio Monteiro de Carvalho, PUC-RJ, Brazil
Marco Antônio Casanova, PUC-RJ, Brazil
Marcus Vinicius Alvim Andrade, UFV, Brazil
Maria Cecília Calani Baranauskas, Unicamp, Brazil
Mario J. Silva, Universidade de Lisboa, Portugal
Max Egenhofer, University of Maine, USA
Miguel Torres, Centro de Investigacion em Computacion, Mexico
Paulo Justiniano Ribeiro Jr., UFPR, Brazil
Renato Assunção, UFMG, Brazil
Ricardo Ciferri, UFSCar, Brazil

Ricardo da Silva Torres, Unicamp, Brazil
Sergei Levashkin, Centro de Investigacion em Computacion, Mexico
Shashi Shekhar, University of Minnesota, USA
Stephan Winter, University of Melbourne, Australia
Valéria Soares, UFRN, Brazil
Valeria Times, UFPE, Brazil
Werner Kuhn, University of Münster, Germany
Yola Georgiadou, ITC, Netherlands

Contents

Distributed GIS / GIS and the Internet

List of Contributing Authors

Adler Cardoso Gomes da Silva
Departamento de Engenharia de Computação e Automação Industrial, Faculdade de Engenharia Elétrica e de Computação, Universidade Estadual de Campinas
Brazil
acardoso@dca.fee.unicamp.br

Angelo Augusto Frozza
Departamento de Ciências Exatas e Tecnológicas, Universidade do Planalto Catarinense
Brazil
frozza@uniplac.net

Antônio Miguel Vieira Monteiro
Divisão de Processamento de Imagens, Instituto Nacional de Pesquisas Espaciais
Brazil
miguel@dpi.inpe.br

Antônio Ramalho-Filho
Centro Nacional de Pesquisa de Solos, Empresa Brasileira de Pesquisa Agropecuária
Brazil
ramalho@cnps.embrapa.br

Antonio Valério Netto
Cientistas Associados
Brazil
antonio.valerio@cientistasassociados.com.br

Bernhard Mitschang
Institute of Parallel and Distributed Systems, Universität Stuttgart
Germany
bernhard.mitschang@ipvs.uni-stuttgart.de

Cirano Iochpe
Instituto de Informática, Universidade Federal do Rio Grande do Sul / Empresa de Tecnologia da Informação e Comunicação da Prefeitura de Porto Alegre
Brazil
ciochpe@inf.ufrgs.br

Cláudio de Souza Baptista
Departamento de Sistemas e Computação, Universidade Federal de Campina Grande
Brazil
baptista@dsc.ufcg.edu.br

Clodoveu Augusto Davis Jr.
Instituto de Informática, Pontifícia Universidade Católica de Minas Gerais
Brazil
clodoveu@pucminas.br

Daniela Francisco Brauner
Departamento de Informática, Pontifícia Universidade Católica do Rio de Janeiro
Brazil
dani@inf.puc-rio.br

Daniela Nicklas
Institute of Parallel and Distributed Systems, Universität Stuttgart
Germany
daniela.nicklas@ipvs.uni-stuttgart.de

Danilo Lourenço Lopes
Departamento de Estatística, Universidade Federal de Minas Gerais
Brazil
danilolopes@ufmg.br

Danilo Palomo
Divisão de Processamento de Imagens, Instituto Nacional de Pesquisas Espaciais
Brazil
danilo@dpi.inpe.br

Eduardo de Rezende Francisco
Escola de Administração de Empresas de São Paulo, Fundação Getúlio Vargas / AES Eletropaulo
Brazil
erfrancisco@gvmail.br

Elvis Rodrigues da Silva
Departamento de Sistemas e Computação, Universidade Federal de Campina Grande
Brazil
elvis@dsc.ufcg.edu.br

Fábio Luiz Leite Jr.
Departamento de Sistemas e Computação, Universidade Federal de Campina Grande
Brazil
fabio@dsc.ufcg.edu.br

Felipe Zambaldi
Escola de Administração de Empresas de São Paulo, Fundação Getúlio Vargas
Brazil
zambaldi@gvmail.br

Flávio L. Mello
Departamento de Engenharia Eletrônica e de Computação, Universidade Federal do Rio de Janeiro
Brazil
flavioluis.mello@gmail.com

Francisco Aranha
Escola de Administração de Empresas de São Paulo, Fundação Getúlio Vargas
Brazil
francisco.aranha@fgvsp.br

Geraldo Zimbrão
Departamento de Ciência da Computação, Universidade Federal do Rio de Janeiro
Brazil
zimbrao@cos.ufrj.br

Gilberto Câmara
Divisão de Processamento de Imagens, Instituto Nacional de Pesquisas Espaciais
Brazil
gilberto@dpi.inpe.br

Gilberto Ribeiro de Queiroz
Divisão de Processamento de Imagens, Instituto Nacional de Pesquisas Espaciais
Brazil
gribeiro@dpi.inpe.br

Guillermo Nudelman Hess
Instituto de Informática, Universidade Federal do Rio Grande do Sul / Dipartimento di Informatica e Comunicazione, Università degli Studi di Milano
Brazil / Italy
hess@inf.ufrgs.br

Jano Moreira de Souza
Departamento de Ciência da Computação, Instituto de Matemática, Universidade Federal do Rio de Janeiro
Brazil
jano@cos.ufrj.br

João Luiz Dihl Comba
Instituto de Informática, Universidade Federal do Rio Grande do Sul
Brazil
comba@inf.ufrgs.br

Karla Donato Fook
Departamento Acadêmico de Informática, Centro Federal de Educação Tecnológica do Maranhão
Brazil
karla@dpi.inpe.br

Leonardo Guerreiro Azevedo
Departamento de Ciência da Computação, Universidade Federal do Rio de Janeiro
Brazil
azevedo@cos.ufrj.br

Leonardo Lacerda Alves
Instituto de Informática, Pontifícia Universidade Católica de Minas Gerais
Brazil
leonardo@lacerda.eti.br

Luiz Felipe C. Ferreira da Silva
Seção de Engenharia Cartográfica, Instituto Militar de Engenharia
Brazil
felipe@ime.eb.br

Marcelo Antonio Nero
Cientistas Associados
Brazil
marcelo.nero@cientistasassociados.com.br

Marco Antônio Casanova
Departamento de Informática, Pontifícia Universidade Católica do Rio de Janeiro
Brazil
casanova@inf.puc-rio.br

Marcus Vinícius Alvim Andrade
Departamento de Informática, Universidade Federal de Viçosa
Brazil
marcus@dpi.ufv.br

Margareth Simões Penello Menezes
Departamento de Engenharia de Sistemas e Computação, Universidade do Estado do Rio de Janeiro (UERJ) / Centro Nacional de Pesquisa de Solos, Empresa Brasileira de Pesquisa Agropecuária (Embrapa)
Brazil
maggie@eng.uerj.br, margaret@cnps.embrapa.br

Matthias Grossman
Institute of Parallel and Distributed Systems, Universität Stuttgart
Germany
matthias.grossman@ipvs.uni-stuttgart.de

Miguel Fornari
Instituto de Informática, Universidade Federal do Rio Grande do Sul / Faculdade de Tecnologia, SENAC-RS
Brazil
fornari@ieee.org

Milton R. Ramirez
Instituto de Matemática, Universidade Federal do Rio de Janeiro
Brazil
milton@labma.ufrj.br

Mirella Antunes de Magalhães
Departamento de Informática, Universidade Federal de Viçosa
Brazil
mirella@dpi.ufv.br

Nelkis de la Orden Medina
Cientistas Associados
Brazil
nelkis.medina@cientistasassociados.com.br

Patrício de Alencar Silva
Departamento de Sistemas e Computação, Universidade Federal de Campina Grande
Brazil
patricio@dsc.ufcg.edu.br

Rafael Goldszmidt
Escola de Administração de Empresas de São Paulo, Fundação Getúlio Vargas
Brazil
rafaelgoldszmidt@gvmail.br

Renata Fernandes Viegas
Programa de Pós-Graduação em Sistemas e Informação, Universidade Federal do Rio Grande do Norte
Brazil
renatafviegas@gmail.com

Renato Martins Assunção
Departamento de Estatística, Universidade Federal de Minas Gerais
Brazil
assuncao@est.ufmg.br

Ricardo Luis Guimarães dos Santos
Cientistas Associados
Brazil
ricardo.guimaraes@cientistasassociados.com.br

Rodrigo Amaral Lapa
Cientistas Associados
Brazil
rodrigo.lapa@cientistasassociados.com.br

Rodrigo P. D. Ferraz
Centro Nacional de Pesquisa de Solos, Empresa Brasileira de Pesquisa Agropecuária
Brazil
rodrigo@cnps.embrapa.br

Ronaldo dos Santos Mello
Departamento de Informática e Estatística, Universidade Federal de Santa Catarina
Brazil
ronaldo@inf.ufsc.br

Ruy Luiz Milidiú
Departamento de Informática, Pontifícia Universidade Católica do Rio de Janeiro
Brazil
milidiu@inf.puc-rio.br

Sérgio Souza Costa
Divisão de Processamento de Imagens, Instituto Nacional de Pesquisas Espaciais
Brazil
skosta@gmail.com

Silvana Castano
Dipartimento di Informatica e Comunicazione, Università degli Studi di Milano
Italy
castano@dico.unimi.it

Thaís Correa
Departamento de Estatística, Universidade Federal de Minas Gerais
Brazil
tataest@yahoo.com.br

Thomas Schwarz
Institute of Parallel and Distributed Systems, Universität Stuttgart
Germany
thomas.schwarz@ipvs.uni-stuttgart.de

Valéria Gonçalves Soares
Departamento de Informática, Universidade Federal da Paraíba
Brazil
valeria@di.ufpb.br

Victor Hugo Meirelles Azevedo
Departamento de Engenharia de Sistemas e Computação, Universidade do Estado do Rio de Janeiro
Brazil
vhmeirelles@gmail.com

Vinícius Lopes Rodrigues
Departamento de Informática, Universidade Federal de Viçosa
Brazil
vlopes@dpi.ufv.br

Wladimir S. Meyer
Instituto Alberto Luiz Coimbra de Pós Graduação e Pesquisa de Engenharia, Universidade Federal do Rio de Janeiro
Brazil
wsmeyer@cos.ufrj.br

Wu Shin-Ting
Departamento de Engenharia de Computação e Automação Industrial, Faculdade de Engenharia Elétrica e de Computação, Universidade Estadual de Campinas
Brazil
ting@dca.fee.unicamp.br

Consistent Handling of Linear Features in Polyline Simplification

Adler C. G. da Silva, Shin-Ting Wu

Departamento de Engenharia de Computação e Automação Industrial
Faculdade de Engenharia Elétrica e de Computação
Universidade Estadual de Campinas

1 Introduction

Polyline simplification is one of most thoroughly studied subjects in map generalization. It consists in reducing the number of vertices of a polygonal chain in order to represent them at a smaller scale without unnecessary details. Besides its main application in generalization, it is also considerably employed in Geographic Information Systems (GIS) to reduce digital map data for speeding up processing and visualization and to homogenize different data sets in the process of data integration. A variety of techniques has been presented by researchers in different contexts [14, 7, 12, 16].

In automated cartography, the most used algorithms are the classical Ramer-Douglas-Peucker (RDP) algorithm [10, 3], Visvalingam's algorithm [17] and Wang and Müller's algorithm [18]. Unfortunately, like the majority of algorithms, none of them maintains the spatial relationship among features and, hence, cannot preserve the original topology of most maps (Figure 1). This is because they take the polyline in isolation, without considering the features in its vicinity. Many ideas [9, 4, 8] have been published in attempt to remove the topological conflicts in a post-processing stage, but, in the cases where the original data is not present, some inconsistencies cannot be besided.

There exists a second class of algorithms which takes into consideration the whole map throughout the course of the simplification [15, 1, 16]. In

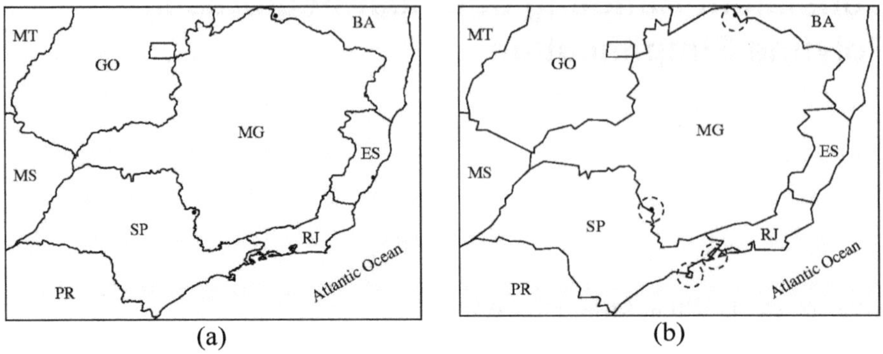

Fig. 1. (a) Original map and (b) its inconsistent simplification outcome.

these techniques, a constrained Delaunay triangulation is performed on the whole collection of map features in a pre-processing stage. The triangulation-based approach permits these algorithms to implicitly preserve the topology of the map while removing vertices from the polylines. Due to its great capability to store spatial relationships and to detect topological conflicts, the Delaunay triangulation is also used to implement other generalization operators, such as exaggeration, collapse and amalgamation [6]. However, when the concern is only on the simplification, this approach may be expensive, since the vicinity of any feature or subfeature must be retriangulated whenever it is removed from the map.

A third class of algorithms modifies a polyline in context, taking into consideration only the relationship between the polyline with nearby features, instead of the complete map features. In these techniques, there is no need for a pre-processing or post-processing stage. The topology of the polyline is conserved along the simplification procedure by preserving the sidedness of the features that are inside its convex hull. Many of these techniques are based on an isolated simplification procedure and simply include the sidedness topological constraint when selecting a vertex to be inserted in or removed from the polyline. This simplification approach may be an alternative to the triangulation-based ones for efficiently solving topological conflicts.

Well-known algorithms on simplification in context are presented in the papers of de Berg *et al.* [2] and Saalfeld [13]. The former works on subdivision simplification, where the polylines are always part of two polygons in the map. It succeeds in generating a topologically consistent polyline that is at a maximum error ε from the original one and has as few vertices as possible. The latter works on a more general polyline simplification, involving linear features that may not be part of a polygon, as, for example, rivers and roads. It improves the classical RDP algorithm for recovering the topology of the original polyline. Saalfeld's algorithm is more popular than de Berg *et al.*'s, because of the popularity of the RDP algorithm, and also because of its simpler implementation and faster processing.

For the sake of simplicity, the algorithms on simplification in context unify the handling of linear and point features by considering the vertices of a linear feature as point features. However, in some particular cases, even if the vertices of a line segment (handled as point features) lie on the correct side, the line segment can still intersect the simplified polyline. The example of Figure 2 illustrates this situation. Before the simplification (Figure 2(a)), the points p_1 and p_2 lie outside the shaded region. After the simplification (Figure 2(b)), although the sidedness of the points is preserved, the line segment p_1p_2 intersects the simplified polyline $\mathcal{P}'$.

Fig. 2. A case where point feature consistency fails in handling linear features: (a) the original polyline $\mathcal{P}$ and the line segment p_1p_2, and (b) the simplified polyline $\mathcal{P}'$ that preserves the sidedness of p_1 and p_2, but intersects p_1p_2.

Our motivation for this work is twofold. Firstly, we would like to remedy the inconsistent outcomes when handling linear features as point features. Secondly, we would like to devise an incremental sidedness test that is appropriate for handling linear features and can be easily integrated to simplification algorithms. The remainder of this paper is organized as follows. We present, in Section 2, a brief analysis of how consistency has been studied in the previous works. Then, in Section 3, we explain how to overcome the sidedness inconsistency along the simplification of linear features. Afterwards, in Section 4, we give an algorithmic solution to Saalfeld's algorithm. After then, in Section 5, we give some basic results of our strategy and compare them to Saalfeld's solution. Finally, in Section 6, we present our concluding remarks and future research directions.

2 Related Work

As previously stated, de Berg *et al.*'s work is on subdivision simplification. They assume that every polyline of a map is part of two polygons of the subdivision. For the purpose of validating their procedure, they formalized the definition of consistency of polylines with respect to point features as follows. Let $\mathcal{P}$ and $\mathcal{P}'$ be two simple polylines oriented from vertex v_1 to vertex v_n, and let F be a set of point features. The polylines $\mathcal{P}$ and $\mathcal{P}'$ are said to be consistent with respect to F, if there exists a simple polyline $\mathcal{C}$ oriented from vertex v_n to vertex v_1 that closes both $\mathcal{P}$ and $\mathcal{P}'$ to simple polygons which have the same subset of points of F in their interior as depicted in Figure 3. One can show that if there exists such a polyline $\mathcal{C}$, then any other simple polyline that closes both $\mathcal{P}$ and $\mathcal{P}'$ in simple polygons will give the same result. In other words, the polylines $\mathcal{P}$ and $\mathcal{P}'$ are consistent with respect to F no matter what polygons of the subdivision they are part of.

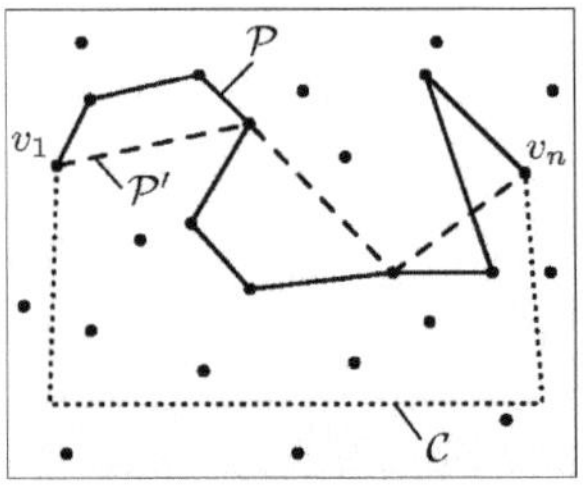

Fig. 3. Consistency of polylines $\mathcal{P}$ and $\mathcal{P}'$ with respect to a set of point features.

The reasoning of this definition is quite simple. Let us consider that, after the simplification of the configuration depicted in Figure 3, the polygon $\mathcal{B}$, formed by $\mathcal{P}$ and $\mathcal{C}$ (Figure 4(a)), was replaced by the polygon $\mathcal{B}'$, formed by $\mathcal{P}'$ and $\mathcal{C}$ (Figure 4(b)). A point feature is consistently placed with respect to $\mathcal{B}$ and $\mathcal{B}'$, if it lies inside or outside both polygons. In Figure 4(b), the point features lying inside $\mathcal{B}$ and outside $\mathcal{B}'$ are indicated with downward arrows and those lying outside $\mathcal{B}$ and inside $\mathcal{B}'$ are indicated with upward arrows. From Figure 4(c), we may see that the point features lying between $\mathcal{P}$ and $\mathcal{P}'$ are the only points that have different sidedness classification with respect to $\mathcal{P}$ and $\mathcal{P}'$.

The definition of consistency given by de Berg *et al.* is valid only for point features. Without additional constraints, the point feature consistency cannot be used for handling linear features. Applying this definition on the cases illustrated in Figure 2, we may close the polyline $\mathcal{P}$ and $\mathcal{P}'$ and build the polygons $\mathcal{B}$ and $\mathcal{B}'$, respectively, as shown in Figure 5. Even with all points on the correct side with respect to $\mathcal{B}$ and $\mathcal{B}'$, intersections still occur. This is because although the extremes of the segment are on the correct side, its intermediate points lie on the wrong side.

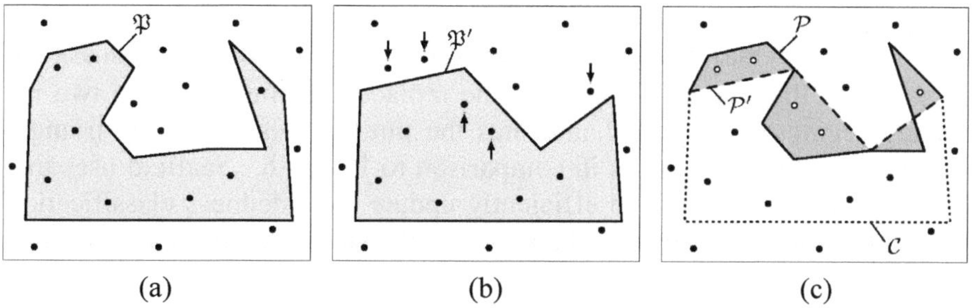

Fig. 4. (a, b) Consistency of the polygons $\mathfrak{P}$ and $\mathfrak{P}'$ with respect to a set of point features. (c) The only features that are on the wrong side with respect to the polygons lie between the polylines $\mathcal{P}$ and $\mathcal{P}'$.

Fig. 5. Case of inconsistency with areal features: (a) original polygon of a subdivision and (b) inconsistent simplified polygon.

In his work, Saalfeld concentrates not only on features that are on the wrong polygon, but also on features that lie on the wrong side of a polylines. According to him, point features always change their sidedness, if they are trapped between $\mathcal{P}$ and $\mathcal{P}'$. Figure 6 gives an example of this situation. Among the point features, only the white ones lie between $\mathcal{P}$ and $\mathcal{P}'$. Three of them are below $\mathcal{P}$ and above $\mathcal{P}'$ and two of them are above $\mathcal{P}$ and below $\mathcal{P}'$. Actually, this is a generalization of the point feature consistency, defined by de Berg *et al.*, for polylines that are not part of polygons.

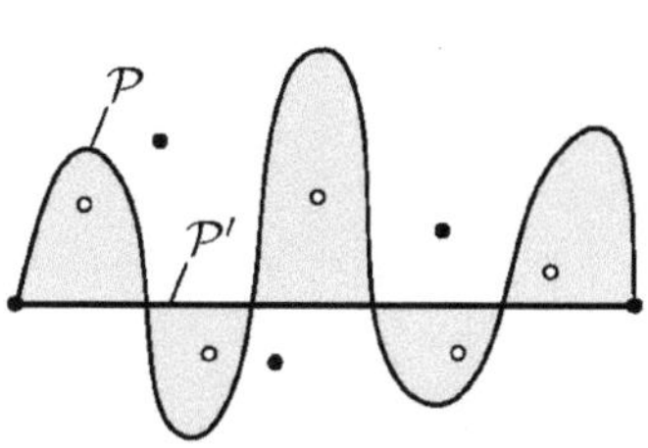

Fig. 6. The sidedness of polylines for detecting inconsistent point features.

Another important contribution of Saalfeld's work is the triangle inversion property, stated as follows. When two segments replace one segment in $\mathcal{P}'$ (or vice-versa), the only point features that invert their sidedness are those inside the triangle formed by the replaced segment and the two replacing segments. Figure 7 indicates the three points inside the triangle that inverted their sidedness in comparison to Figure 6. Saalfeld uses this property in his algorithm to efficiently update the sidedness classification of features after the insertion of a vertex in the simplified polyline.

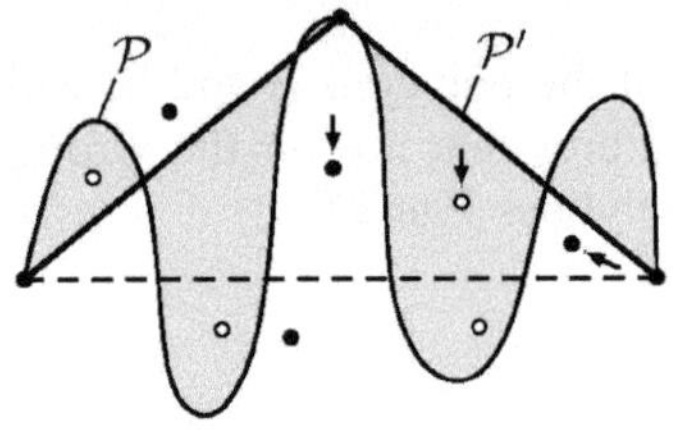

Fig. 7. Triangle inversion property.

So far as we know works on simplification in context based on reducing linear to point features are not able to correctly handle linear features.

3 Handling Linear Features

First of all, we have to identify the cases where the consistency condition for point features fails with linear features. Let us consider the configuration given in Figure 8(a), where the line segments v_iv_k and v_kv_j replace respectively the subpolylines $\mathcal{P}_{ik}$ and $\mathcal{P}_{kj}$. Notice that p_1 is considered to be on the correct side, even if some intermediate points of the line segment p_1p_2 are not. That is because the subpolyline $\mathcal{P}_{kj}$ crosses the line segment v_iv_k and forms the region depicted in Figure 8(b) where p_1 lies. One can show that inconsistencies may occur whenever a subpolyline $\mathcal{P}_{ab}$ crosses the simplifying segment v_cv_d of another subpolyline $\mathcal{P}_{cd}$.

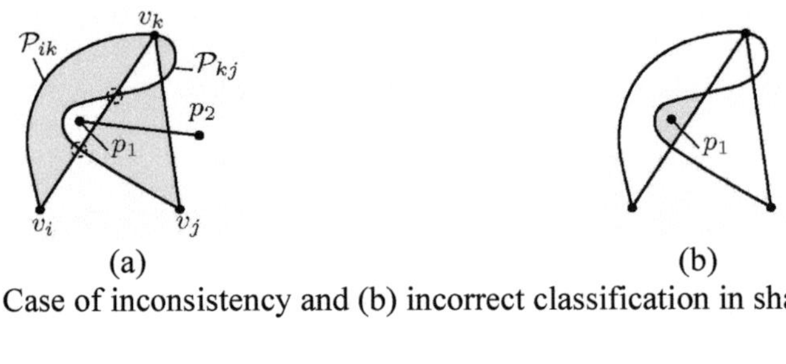

Fig. 8. (a) Case of inconsistency and (b) incorrect classification in shaded region.

The solution we adopted is simple. We apply separately the sidedness criterion to each subpolyline and its simplifying segment. Figures 9(a) and 9(b) show the application of this criterion for the case of Figure 8. Notice that p_1 is on the wrong side with respect to both subpolylines, but with respect to $\mathcal{P}$ and $\mathcal{P}'$ it is on the correct side. We formalize the consistency for linear features as follows. Let $\mathcal{P}$ be a polyline, $\mathcal{P}'$ be a simplified version of $\mathcal{P}$, and F be a set of vertices of linear features. The polylines $\mathcal{P}$ and $\mathcal{P}'$ are said to be consistent with respect to F, if the polygons formed by each subpolyline $\mathcal{P}_{ij}$ and its correspondent line segment v_iv_j contain no element of F. Figure 9(c) illustrates an example of consistent simplification

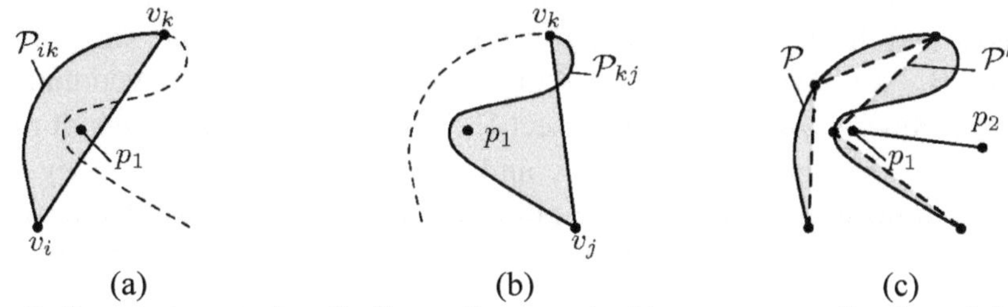

Fig. 9. Our strategy to handle linear features: (a, b) separate application of the sidedness criterion to each subpolyline; (c) an example of consistent simplification.

We consider that the interior of the polygons (represented by the shaded regions in Figure 9) are determined with the parity (or odd-even) rule. We compute the number of crossings between a ray from the feature and the polygon formed by a subpolyline and its correspondent simplifying line segment. If the number of crossings is odd, the feature is on the wrong side; otherwise, it is on the correct side (Figures 10(a) and 10(b)). For elucidating how the linear feature consistency works, we introduce the parity property as follows. Two points are considered to be on the same side of a polygon, if a line connecting them crosses the polygon an even number of times. Otherwise, they are considered to be on opposite sides. Figures 10(c) and 10(d) illustrate two examples of the parity property. Notice that the crossings on self-intersecting points of a polyline are counted as many as the number of segments that intersect on it.

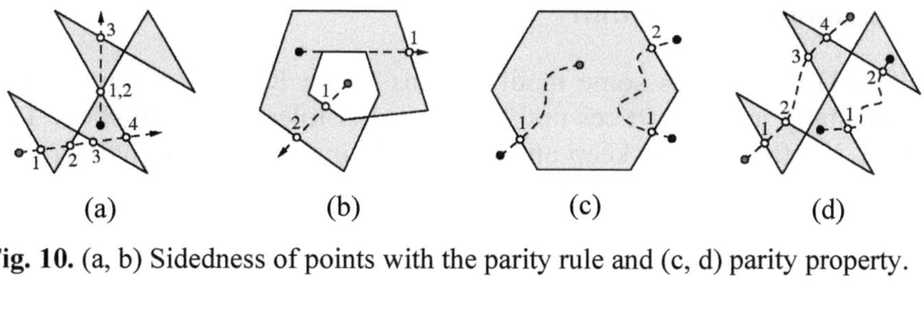

Fig. 10. (a, b) Sidedness of points with the parity rule and (c, d) parity property.

We proceed to formalize a sufficient point feature-based condition for ensuring consistent linear simplification.

Proposition 1. *Let* $\mathcal{P}_{ij}$ *be a subpolyline of the polyline* $\mathcal{P}$ *and let* v_iv_j *be its correspondent simplifying segment. If a line segment* p_1p_2 *does not intersect the original polyline* $\mathcal{P}$*, and the points* p_1 *and* p_2 *are both outside the polygon* $\mathcal{B}_{ij}$ *formed by* $\mathcal{P}_{ij}$ *and* v_iv_j*, then* p_1p_2 *does not intersect* v_iv_j*.*

Proof. From the fact that p_1p_2 does not intersect $\mathcal{P}$ (and consequently $\mathcal{P}_{ij}$) and p_1 and p_2 are both outside $\mathcal{B}_{ij}$, we have that p_1 and p_2 do not coincide with the line segment v_iv_j, and, consequently, p_1p_2 and v_iv_j do not overlap. Therefore, they can intersect at most in one single point. From the parity property and from the fact that p_1 and p_2 are on the same side of the polygon $\mathcal{B}_{ij}$, we have that the line segment p_1p_2 crosses $\mathcal{B}_{ij}$ an even number of times. Since p_1p_2 does not intersect $\mathcal{P}_{ij}$, if there is any crossings, it must be between the line segments v_iv_j and p_1p_2. However, since they can intersect in no more than a single point, the number of crossings between p_1p_2 and $\mathcal{B}_{ij}$ to be even must be zero. Hence, p_1p_2 does not intersect v_iv_j. □

When applying the conditions of Proposition 1 to each subpolyline of $\mathcal{P}$ and its correspondent line segment in $\mathcal{P}'$, we ensure that p_1p_2 will not intersect p_1p_2. Hence, our approach guarantees that any linear feature that does not intersect the original polyline $\mathcal{P}$ will not intersect the segments of the simplified polylines $\mathcal{P}'$. Since it is more restrictive than the point feature consistency, we can use it to uniformly handle both point and linear features, without making any distinction between them.

4 The Algorithm

In this section we present an algorithmic solution for correctly handling linear features. We replace the triangle inversion test devised by Saalfeld by our proposed strategy. To be self-contained, Saalfeld's algorithm is briefly presented in Section 4.1 and then our solution is described in Section 4.2.

4.1 Saalfeld's Algorithm

Saalfeld [13] proposes some modifications to the RDP algorithm, which incrementally inserts vertices until a given error tolerance is satisfied. His strategy is, after then, to keep on successively inserting additional vertices to the "inconsistent" segments of the polyline until all errors are removed. As illustrates the flowchart in Figure 11, his procedure is convergent, because, in the worst case, it adds all vertices of the original polyline, recovers the original geometry and, consequently, the original topology of the map. Naturally, in real data sets, the worst case rarely occurs. His algorithm is divided in two steps. The first step consists only in the application of the RDP in the input polyline $\mathcal{P}$ and the second step is comprised of the topological correction procedures.

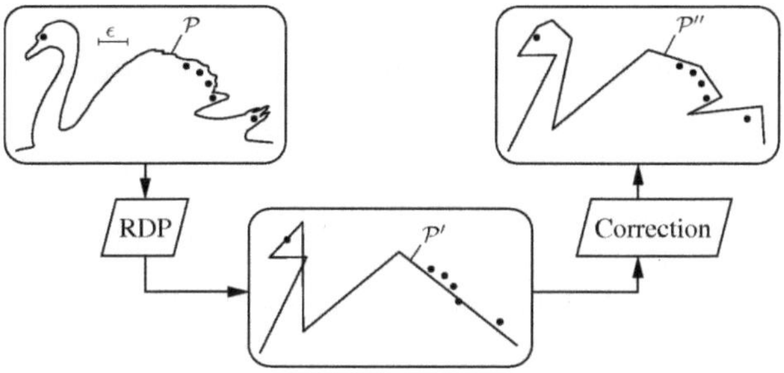

Fig. 11. Flowchart of Saalfeld's algorithm.

The correction step is depicted in Figure 12 and works as follows. For each subpolyline $\mathcal{P}_{ij}$ replaced by the polyline segment v_iv_j in the simplified polyline $\mathcal{P}'$, the algorithm determines its convex hull. Each subpolyline is then associated to the list of features that are inside its convex hull. These features represent potential topological conflicts. This list may include point features, vertices of neighbouring polylines, and the remaining vertices of the polyline itself (namely the vertices v_k such that $k < i$ or $k > j$. The sidedness of these features is computed with the parity rule.

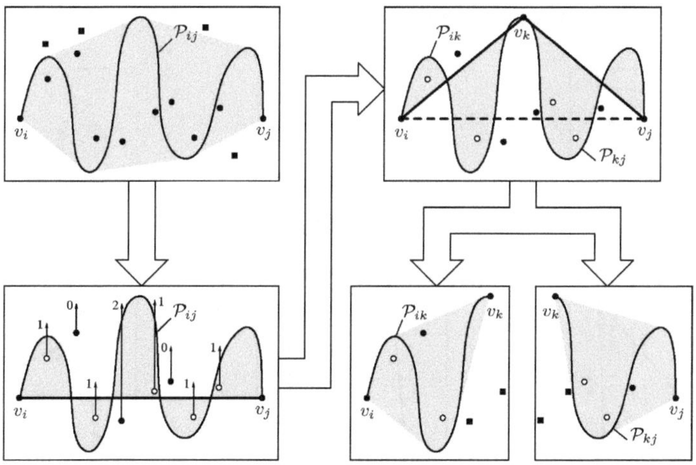

Fig. 12. Correction step of Saalfeld's algorithm: selecting features inside the convex hull, and computing and updating their sidedness classification.

After the initialization, for each subpolyline $\mathcal{P}_{ij}$ that has features on the wrong side, the algorithm breaks its correspondent line segment v_iv_j by adding the farthest vertex v_k. It updates the sidedness classification of the external points in the current convex hull, using the triangle inversion property. Then, it splits the convex hull in two and selects the external points of the resulting convex hulls. After that, it calls the correcting procedure for the subpolylines $\mathcal{P}_{ik}$ and $\mathcal{P}_{kj}$. Since the whole process is restarted independently for the two subpolylines, the vertices of $\mathcal{P}_{ik}$ must be checked with respect to the convex hull of $\mathcal{P}_{kj}$, and vice-versa. If some vertices of one subpolyline interfere in the other subpolyline convex hull, their sidedness is computed with the parity rule.

Let us consider the application of Saalfeld's algorithm under tolerance ∞ to the polyline $\mathcal{P}$ of Figure 13(a). Because of the ∞-tolerance, the first step (RDP algorithm) adds no vertices to $\mathcal{P}'$. In the second step the algorithm first calculates the number of crossings of the points p_1, p_2, and p_3 and classifies their sidedness (Figure 13(b)). Since p_3 is on the wrong side, it adds the farthest vertex v_4 and updates the sidedness classification of p_3, which is inside the triangle $\Delta v_1v_4v_8$ (Figure 13(c)). Then, it handles independently the subpolylines $\mathcal{P}_{1,4}$ and $\mathcal{P}_{4,8}$, after evaluating the dependency of their vertices. As all the features are on the correct side with respect to $\mathcal{P}_{4,8}$, no further splitting should be applied on it. Regarding to $\mathcal{P}_{1,4}$, the vertices v_5 and v_6 are inserted in its list of features (Figure 13(d)). Both vertices are considered to be on the wrong side. The algorithm adds the vertex v_3 and updates the sidedness classification of p_1, v_6 and v_7 (Figure 13(e)). Observe that p_1 is on the wrong side, but the algorithms stops. That is because there are no more changes for $\mathcal{P}_{1,3}$ and $\mathcal{P}_{3,4}$.

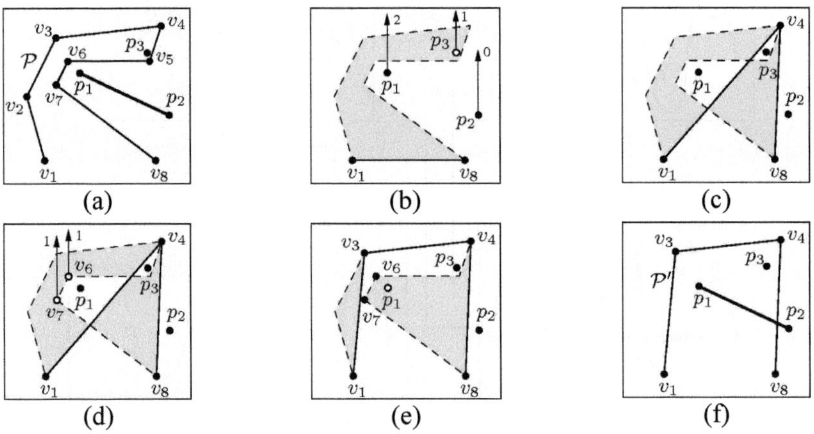

Fig. 13. Failure in Saalfeld's algorithm.

Observe, from Figure 13(c), that the triangle inversion property is equivalent to the point consistency strategy of de Berg *et al.*. However, Saalfeld's algorithm adds more vertices to the simplified polyline, due to its independent treatment of distinct subpolylines. Nevertheless, his algorithm is not yet able to remove all intersections.

4.2 Update of Sidedness Classification

Our strategy is based on the fact that the relationship between a feature and the original polyline never changes throughout the course of the simplification. This permits us to associate the feature to precomputed data that store this relationship. We divide our strategy in two stages. In the first stage, we compute the crossings between the upward ray from a feature f and a subpolyline $\mathcal{P}_{ij}$, as depicted in Figure 14(a), and store them in a data structure associated to f. In the second stage, after breaking the line segment v_iv_j in two new line segments v_iv_k and v_kv_j, we update the sidedness classification of f with respect to the subpolylines $\mathcal{P}_{ik}$ and $\mathcal{P}_{kj}$, as shown in Figures 14(b) and 14(c), respectively. We present a pseudocode of our algorithm that can replace the triangle inversion test.

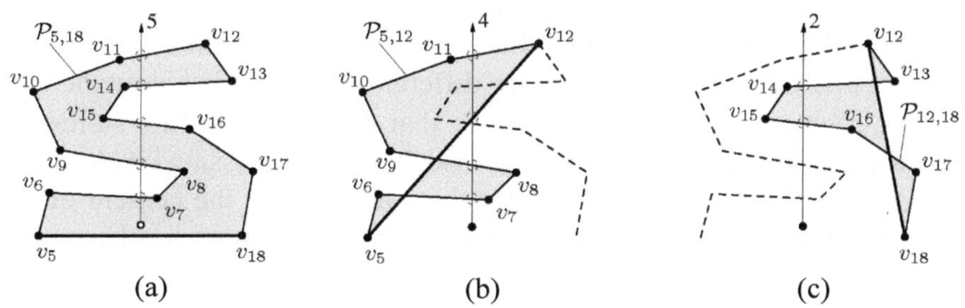

Fig. 14. Parity test: (a) computation of crossings between the upward ray from feature f and the subpolyline $\mathcal{P}_{5,18}$; (b, c) update of its sidedness classification with respect to the resulting subpolylines.

Besides its point coordinates `x` and `y`, we associate to the feature f the array `crossings` and the indices `begin` and `end`, as depicted in the structure of Figure 15(a). The array `crossings` is used to store the indices of the line segments of $\mathcal{P}_{ij}$ that cross the upward ray from f. Since in real maps the number of crossings is usually very small in comparison to the number of line segments being processed, one expects the array `crossings` to be very small too. The variables `begin` and `end` store initially the first and the last indices of `crossings`. Figure 15(b) illustrates the initial state of the array `crossings` and the indices `begin` and `end`. The numbers stored in the fields of `crossings` corresponds to the indices j of the segments $v_i v_j$ that the upward ray intersects. Observe that the number of crossings can be directly obtained by the subtraction (`end-begin`).

```
Type Feature
Begin
   x, y: Float;
   crossings: Integer[];
   begin, end: Integer;
End
```

(a)

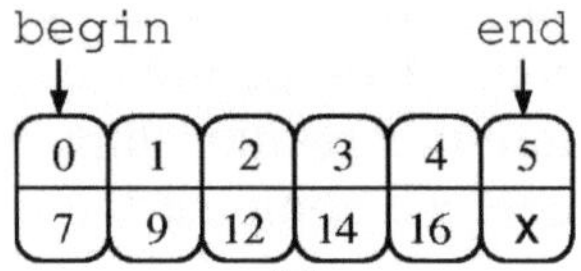

(b)

Fig. 15. The data structure used in our strategy (a) in pseudocode and (b) its graphical representation for the polyline of Figure 14(a).

The procedure `computeCrossings` outlined in Code 1 performs the computation of the crossings between the ray and the original subpolyline, stores them in the array and initializes the indices. From line 08, a crossing is found when (1) the subpolyline changes its side with respect to the ray (`currSide` and `prevSide` are different) and (2) f is below the current line segment. Observe, in line 09, that the algorithm first stores the indices in a linked list. This is because, before processing, it does not know the number of crossings. After the computation, the content of the list is finally copied to the array, for which enough memory has been allocated (line 13). Since the search for crossings is done from i+1 to j (line 06), the indices of the crossed line segments are stored in ascending order. The ordered array has a specific purpose in the update stage.

In the update stage, to determine the number of crossings of the upward ray from f with the subpolylines P_{ik} and P_{kj}, the algorithm adopts the following strategy. Since the algorithm has already computed the crossings with P_{ij} and stored them, it just looks for the first element after the index k (of the breaking vertex v_k) in the array `crossings`. Let us reconsider the example of Figure 14, where k = 12. The algorithm allocates two distinct copies of f, one for P_{ik} and another for P_{kj}. Then, it looks for the first element in `crossings` greater than k, which is the element 14 of index 3. After then, it updates the index `begin` and `end` of the two copies of f as depicted in Figure 16. The ascending order of the array `crossings` permits the algorithm to perform a binary search. To avoid the overhead of copying `crossings`, the copies of f just keep a reference to it. After the update process, we can obtain the number of crossings for each subpolyline just by subtracting the new indices.

```
Procedure computeCrossings(P: Polyline; i, j: Integer; var p: Feature)
Var
   k: Integer;
   list: IntegerList;
   prevSide, currSide: Side;
Begin
   prevSide ← (P[i].x ≤ p.x) ? left : right;
   For k ← i+1 to j do
      currSide ← (P[k].x ≤ p.x) ? left : right;
      If currSide ≠ prevSide .and. p is below line segment P[k]P[k-1] then
         Push k in list;
      End if
      prevSide ← currSide;
   End for
   Allocate memory for p.crossings and copy the content of list to it;
   p.begin ← 0
   p.end ← list.size
End
```

Code 1. Algorithm for computing the crossings between the upward ray from feature *f* and the subpolyline P_{ij}, and storing the indices of the intersecting segments in the array `crossings` associated to *f*.

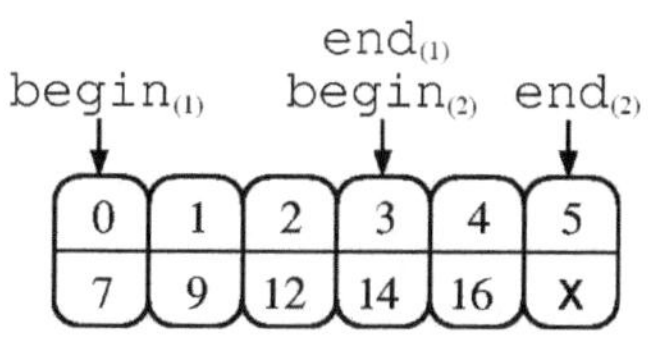

Fig. 16. Resulting indices for the subpolylines in Figures 14(b) and 14(c).

For computing the sidedness of a given feature *f*, the algorithm has time complexity $O(n)$, due to the search for crossings, and memory complexity $O(n)$, due to the array `crossings`. For updating the sidedness classification of *f*, the binary search gives time complexity $O(\log n)$. The new copies of *f* keep just a reference to `crossings`, so there is no overhead for copying the array. The processing time of this algorithm is comparable to the time complexity of the triangle inversion test, because the array of crossings is usually very small.

5 Results

To validate our theoretical study of consistency for linear features, we present some results of our approach and compare them to those obtained by Saalfeld's algorithm. We examine the basic cases where Saalfeld's strategy fails in preserving the original polyline topology. For each image, the first square presents the original polyline, the second square shows the outline of Saalfeld's algorithm, and the third square exhibits the outline of our approach. The results represent typical cases of intersections (Figures 17(a) and 17(b)), self-intersections (Figures 17(c) and 17(d)), and misplaced point features (Figures 17(e) and 17(f)) that occurs in Saalfeld's algorithm and are correctly handle by our procedure.

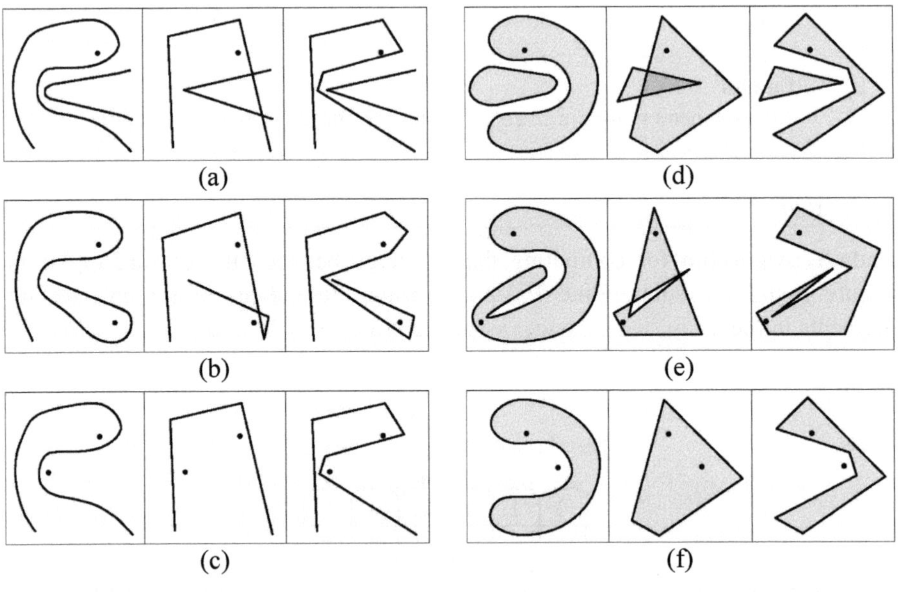

Fig. 17. Comparison between Saalfeld's algorithm and our proposal.

To evaluate the time performance of our proposal with respect to the triangle inversion test in practical cases, we measure their processing times for a variety of data sets with distinct number of points (source [10]). Notice that the processing times of the array of crossings are very close to the ones of the triangle inversion test, even for the map with more points. This is because the number of crossings is very small when compared to the number of vertices of the polylines. Thus, the performance of the algorithms can be considered equivalent. An important result that we achieved with our proposal is that the additional number of vertices that is required to preserve the topological consistency is insignificant. In all cases that we tested, it is less than one vertex in 4,000 inserted by Saalfeld's algorithm (i.e., approximately 0.025%).

Table 1. Comparison of time performance between the triangle inversion test and the strategy with array of crossings.

Original map	Simplified map			
	Triangle inversion		Array of crossings	
#Points	Time [s]	#Points	Time [s]	#Points
26,536	0.930	7,288	0.935	7,288
52,639	1.310	12,165	1.320	12,167
68,506	2.312	15,683	2.330	15,685
103,450	3.010	29,713	3.110	29,713
126,404	3.080	25,769	3.140	25,775
166,157	5.650	24,954	5.750	24,987

6 Concluding Remarks

In this paper, we firstly studied the common problem of using the point feature consistency for handling linear features in topologically consistent polyline simplification algorithms. We observed that, if no pre-processing is carried out in order to satisfy some conditions, a few arrangements of linear features can still lead to intersections. To overcome this problem, we presented a more restrictive consistency constraint for avoiding both changes of sidedness of point and linear features and intersections between linear features. We consider that the sidedness of a point or a vertex of a line segment must be individually checked against each subpolyline $\mathcal{P}_{ij}$ of the original polyline and its correspondent simplifying line segment v_iv_j in the output polyline. This simple theoretical solution permits us to uniformly handle both point and linear features.

In the practical context, the main contribution of this paper lies on an algorithmic strategy that can replace the triangle inversion test employed in Saalfeld's algorithm. Our strategy is based on the fact that, once the crossings are computed, they can be stored in a data structure and recovered whenever one needs. We discussed the ins and outs of the presented strategy and showed that its time complexity is comparable to the triangle inversion test. We also gave a pseudocode of the procedure that may be directly integrated to simplification algorithms. Finally, we presented some results of our procedure and compared it to the ones of triangle inversion test, showing that the procedures have equivalent performances, but our technique always preserves the topology of the original map.

Our future researches point mainly to the development of a topologically consistent simplification procedure that would treat all the polylines together in a global approach. This strategy has the advantage of testing only the current vertices on the simplified polylines, instead of checking all the vertices of the nearby polylines, resulting in better generalizations and faster processing. We intend to separate the simplification procedure from the topological control, so that it may be possible to ensure topological consistency in distinct isolated simplification algorithms. We also plan to place this new algorithm as a faster alternative to the triangulation-based approach for preserving topological consistency in simplification.

Acknowledgments

We would like to acknowledge the Coordination for the Improvement of Higher Education Personnel Foundation (CAPES) and the State of São Paulo Research Foundation (FAPESP, Grant Nº 2003/13090-6) for financial support.

References

[1] Ai, T., Guo, R., & Liu, Y. (2000). A Binary Tree Representation of Curve Hierarchical Structure Based on Gestalt Principles. *The 9th International Symposium on Spatial Data Handling, 2a*, pp. 30-43. Beijing, China.

[2] de Berg, M., van Kreveld, M., & Schirra, S. (1998). Topologically Correct Subdivision Simplification Using the Bandwidth Criterion. *Cartography and Geographic Information Systems, 25* (4), 243-257.

[3] Douglas, D. H., & Peucker, T. K. (1973). Algorithms for the reduction of the number of points required for represent a digitized line or its caricature. *Canadian Cartographer, 10* (2), 112-122.

[4] Edwardes, A., Mackaness, W., & Urvin, T. (1998). Self Evaluating Generalization Algorithms to Automatically Derive Multi Scale Boundary Sets. *The 8th International Symposium on Spatial Data Handling*, (pp. 361-372). Vancouver, Canada.

[5] Jenks, G. F. (1981). Lines, Computers and Human Frailties. *Annals of the Association of American Geographers, 71*, pp. 1-10.

[6] Jones, C. B., Bundy, G. L., & Ware, J. M. (1995). Map generalization with a triangulated data structure. *Cartography and Geographic Information Systems, 22* (4), 317-331.

[7] Lang, T. (1969). Rules for the robot draughtsmen. *The Geographical Magazine, 42* (1), 50-51.
[8] McKeown, D., McMahil, J., & Caldwell, D. (1999). The use of spatial context in linear feature simplification. *GeoComputation 99.* Mary Washington College, Fredericksburg, Virginia.
[9] Müller, J. C. (1990). The removal of spatial conflicts in line generalisation. *Cartography and Geographic Information Systems, 17* (2), 141-149.
[10] Penn State University. (1992). Fonte: Digital Chart of the World Server: http://www.maproom.psu.edu/dcw
[11] Ramer, U. (1972). An iterative procedure for the polygonal approximation of plane curves. *Computer Graphics and Image Processing, 1*, 224-256.
[12] Reumann, K., & Witkam, A. P. (1974). Optimizing curve segmentation in computer graphics. In: A. Gunther, B. Levrat, & H. Lipps (Ed.), *Proceedings of the International Computing Symposium* (pp. 467-472). American Elsevier.
[13] Saalfeld, A. (1999). Topologically consistent line simplification with the Douglas-Peucker algorithm. *Cartography and Geographic Information Science, 26* (1), 7-18.
[14] Tobler, W. R. (1964). *An experiment in the computer generalization of map.* Office of Naval Research, Geography Branch.
[15] van der Poorten, P. M., & Jones, C. B. (2002). Characterisation and generalisation of cartographic lines using Delaunay triangulation. *International Journal of Geographical Information Science, 16* (8), 773-795.
[16] van der Poorten, P. M., & Jones, C. B. (1999). Customisable line generalisation using Delaunay triangulation. *The 19th International Cartographic Association Conference.*
[17] Visvalingam, M., & Whyatt, J. D. (1993). Line Generalisation by Repeated Elimination of Points. *Cartographic Journal, 30* (1), 46-51.
[18] Wang, Z., & Müller, J. C. (1998). Line Generalization Based on Analysis of Shape Characteristics. *Cartography and Geographic Information Systems, 22* (4), 264-275.

Cartographical Data Treatment Analysis for Real Time Three-Dimensional Terrain Visualization

Flávio L. Mello[1], Luiz Felipe C. Ferreira da Silva[2]

[1]Departamento de Engenharia Eletrônica e de Computação
Universidade Federal do Rio de Janeiro
[2]Seção de Engenharia Cartográfica
Instituto Militar de Engenharia

1 Introduction

This article presents an alternative solution for the triangulation problem of altimetric data of cartographic features. This triangulation corresponds to one of the phases of the methodology to be presented. The proposed mechanism helps to enhance the performance of the graphic hardware by reducing the number of primitives to be processed during the exploration of the environment. The main goal of this method is to allow data management obtains real time three-dimensional (3D) terrain visualization from 2D topographic charts. This application is supposed to be executed on low performance computers embedded on military tanks and vehicles. Thus, the methodology is constrained by this obsolete hardware.

The goal of the altimetric data triangulation is to generate a 3D terrain mesh that is associated to a given region. This mesh, composed by triangles geometrically organized, provides the 3D terrain aspect. Besides this conceptual representation, there is a technological constrain related to the number of triangles used for the same representation. The relation between the terrain levels of detail, and its associated data, can easily exceed the video board processing capabilities, which usually makes impossible a real time application. Thus, the terrain models data must be processed in order to allow its usage.

Under an interactive three-dimensional environment, the animation becomes an important characteristic. It is supposed to be continuous and smooth, that is, the user must not notice the frame transition between different images. In computer systems, the responsibility for the scenes transition delay is given by the quantity of objects on the scene [14]. If this scene contains many objects, the video board will take too long to reproduce them on computer screen, delaying the next scene visualization.

Consequently, assuming one video board model, the terrain flyover interactivity, or any other user movement over the terrain, depends indirectly on the quantity of triangles used on the scene representation. In order to ensure an acceptable screen refresh rate, a strategic terrain vision (wide area of interest) may be represented using low levels of detail, while a tactical vision (narrow area of interest) can be represented by a greater quantity of data. The computer refresh rate is kept under acceptable performance according to system requirements by combining the width of the region of interest and its resolution degree. The term "acceptable performance" does not describe deterministically this measure feature. However, it is used so because the performance values vary according to the video board used on the system visualization. Therefore, it is important to understand the real time concept; to define evaluation quality parameters; and to evaluate the cartographic data pre-processing methodology according to several video boards.

2 Real Time Systems

The altimetry pre-processing main goal is to structure data to meet the 3D visualization application needs. In this article, the refresh rate of the scene drawing will be the metric used for analyzing the system performance. These measured values are associated to system ability to handle real time operations, and for this reason it is important to comprehend the systems dedicated to this kind of application.

A system is a connected element set which can be considered as a unique component [12]. A process corresponds to a time sequence state that a system assumes, according to the transformations that take place inside of it [9]. Thus, a scheme managing of a given process can be considered as a real time system if it's processing time, and its corresponding feedbacks, are compatible with the process dynamics [1].

Consequently, the characterization of a real time system environment is related to its execution time, where each application has its own constrains. A real time system label is usually attributed to those responsible for supervising and controlling time response systems, such as industrial plants, electrical and atomic plants, and intensive therapy units.

On the other hand, one could consider systems like maritime traffic control, or cosmic probes navigation. At the first case, a possible unit measure might be nautical miles per hour, while at the second case, it might be used light years per year. By superficially analyzing this study cases, one might consider strange to classify those systems as real time environments due to its elastic time response. However, this is not a sustainable argument because the real time environment definition is not associated to the elapsed time, but it is related to an opportune problem treatment. According to Martin [15], real time systems might have more elastic response time if the answer delay does not compromise the application correct functioning.

In the military environment, a terrain three-dimensional visualization system must also allow real time answers. In the Brazilian Army, the equipment project for military systems [5] is defined by the Basic Operational Specifications that establish the desired characteristics to be attended by the system. Usually, the specifications demands oppose the commercial off-the-shelf technological restrictions like weight limitation, batteries autonomy and processing capacity. For example, the current military batteries can provide energy for displays, radios, electronic sensors and other equipment for a continued 12 hours mission [11]. Fast Action Troops, such as a Parachuting Brigade or an Air Transportable Brigade, need electric energy for at least 72 continuous hours [6]. According to Future Force Warrior [7,8], a United States Army research project, a soldier equipped with wireless communication helmet, night vision camera, PDA (Personal Digital Assistant) and GPS (Global Positioning System) would need to carry 15 kilograms extra combat equipment just for the batteries.

In order to reduce the equipment weight, soldiers can use lighter materials, use new technologies, or even eliminate superfluity. However, the batteries are necessary heavy elements, which cannot be simply discarded. The current problem solution adopted by military forces has been the reduction of the quantity of batteries. In order to preserve the system autonomy time it is necessary to reduce the electronic components energy consumption. This energy cutback is made possible by decreasing the system processing capacity.

In this context, the development of applications for three-dimensional terrain visualization becomes a challenge. First, because the nature of this kind of application involves a great quantity of data to be processed. Second, because the available computational resources tend to be exiguous. The direct implication of these two factors is that the application becomes incapable to provide data in real time. Therefore, it is recommended to smartly prepare the cartographical data before its use by the visualization system, as described in the following sections.

3 The Terrain Mesh

A mesh is a discrete representation from a geometric domain using primitive shapes. In Figure 1, the geometric domain represented by a real terrain model is composed by triangular geometric primitives. Each triangle vertex corresponds to a measurement point on the object surface.

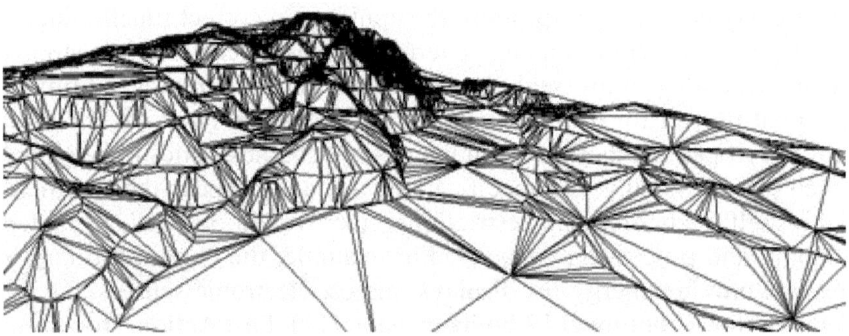

Fig. 1. A terrain model represented by a triangular mesh.

A set point triangulation consists on finding straight line segments that connect these points so that: (1) none of these segments crosses with another straight line segment, and; (2) each point is vertex of at least a triangle formed by these segments (or a minimum of two triangles when considering the convex hull). Thus, to compute the Figure 1 mesh, it is necessary to work out the surface head point's triangulation. This triangulation resulted in the collection of triangles observed in the same figure.

Ideally, to construct a 3D terrain mesh it must be measured all the surface points' heights and map them on the mesh. However, it is impossible to know exactly the heights of all points on a terrestrial surface, that is, it is only possible to know those points' altitude measured on the terrain. This means that it is only feasible to know the correct height of a finite point set. The other surface points' heights must be defined by interpolating those sample heights. An ordinary approach is to attribute the same height to all neighboring points near a sample height. However, as observed in the Figure 2a, the result is quite artificial. On the other hand, another strategy could be described as follows. Initially, each quoted point is connected using straight line segments in order to compose a triangle set. Then, each triangle vertex is raised until its corresponding height, providing a better surface representation at three-dimensional space, as illustrated in Figure 2b [4].

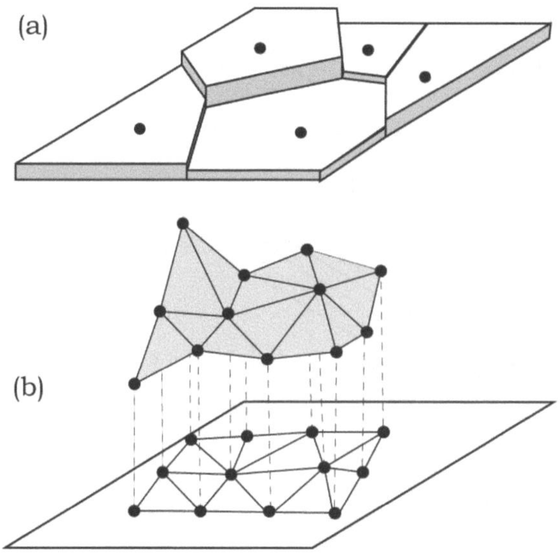

Fig. 2. Terrain surface representation: (a) ordinary solution; (b) less artificial solution using triangles primitives (adapted from Berg et al. [4]).

It must be observed that, given a point set, it is impossible to be certain about the terrain shape, but just the height of the quoted points. Thus, since no other information is provided, it is correct to state that all possible triangulations are equally good. However, some triangulations seem more natural than others. Usually, meshes with strong acute triangles seem to be visually more pleasant than the ones with less acute angles. As a result, the visually better meshes are those that maximize the minimum triangle angles, and known by Delaunay triangulations [10, 19]. An important characteristic from Delaunay triangulation is that, for each sample set, there is a unique triangulation associated to it.

4 The Cone Vision

As described in Mello & Ferreira [17], the cartographic categories are features layers used for structuring digital map libraries, as illustrated in Figure 3. The DSG (Brazilian Geographic Service Directory), for example, implements nine categories: transport system, infrastructure, limits, constructions, control points, hydrography, localities, altimetry and vegetation. Considering a given region, there is a specific toponymy archive associated to each category, totalizing eighteen archives. So, an area mapped by DSG has nine features archives and nine toponymy archives.

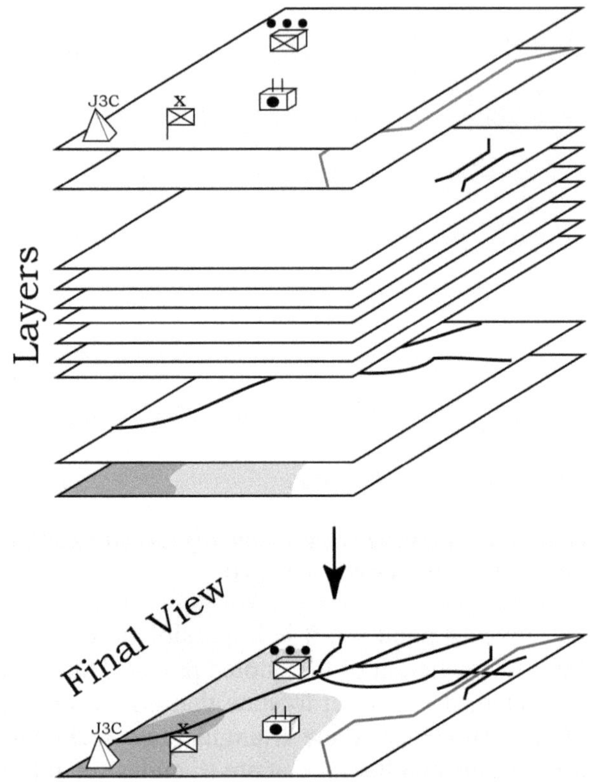

Fig. 3. The representation of any region over a geographic space is composed by overlapping information layers of called cartographic categories.

The quantity of geometric elements available in each cartographic category can become computationally onerous, either under the graphical processing point of view, or under memory consumption. In both cases, the final application will become slower. It is easy to notice that just a fraction of geometries of the virtual environment is effectively visible at a given moment [16], as illustrated in Figure 4. It must be observed that in the instant described in the figure, the soldier can only see the tank and a road stretch. The helicopter is not visible in this situation. In order to reproduce on computers the image observed by the soldier, the video board would not need to draw the whole scene. It would only be necessary to display the tank and the road objects. In this situation, the helicopter would be discarded from the exhibition. Hence, if it is possible to determine the invisible geometries before being drawn, it would be possible simply to ignore them, thus preventing the unnecessary data processing for the application. This clipping procedure is known as frustum culling [3].

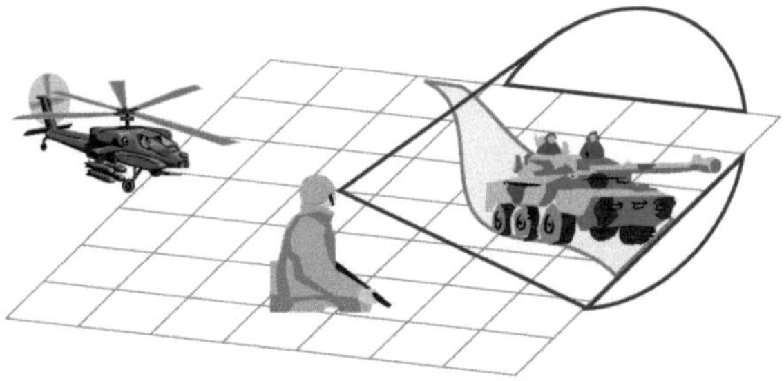

Fig. 4. Object visibility at a given instant: the road and the tank are visible under the soldier point of view, but the helicopter is not.

The virtual environment visible geometries are defined as the intersection between the observer vision cone and the cartographic categories. For the next study case, consider only the altimetry layer of a given region. The common region between the cone vision and the data layer is illustrated in the Figure 5a. In computerized systems it is common to represent the observer vision cone as vision pyramid [13], as presented in the Figure 5b. The overall computations related to pyramids vision tend to be significantly faster than the calculations associated to cones vision, and for this reason, the pyramids vision are used the most.

The region of interest is divided into an indexed cells space [17] in order to allow a unique identification of the areas intercepted by the pyramid vision. This indexation is illustrated in the Figure 5c. Finally, an intersection test is performed with the intention of determining cell totally, or partially, contained in the pyramid vision. Considering the case described on Figure 5d, it can be observed that cells 0-4-5-10-15 had been discarded during the frustum culling procedure, meaning that all the geometries contained in its interior would not be sent for the video board, avoiding unnecessary computer video processing.

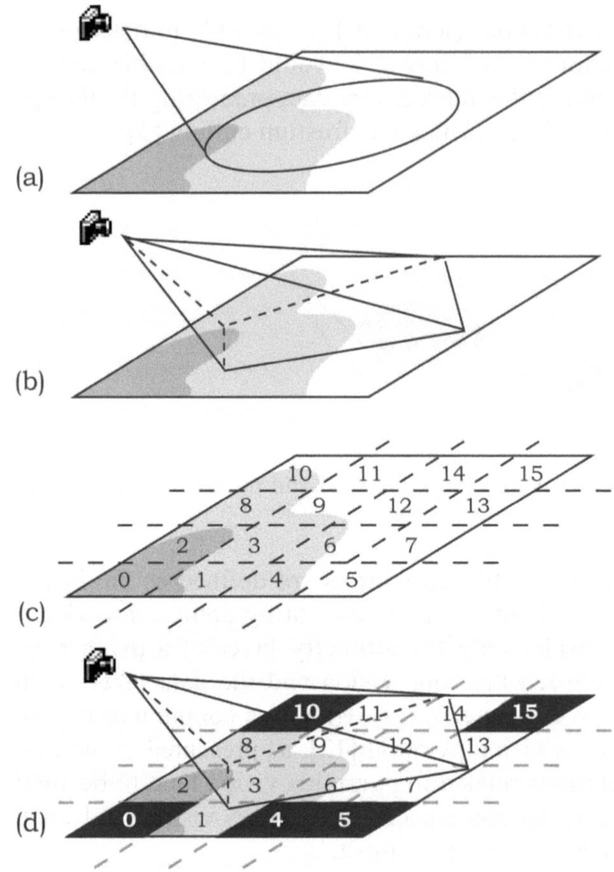

Fig. 5. (a) A vision cone delimiting the visible area of an altimetry layer; (b) A pyramid vision delimiting the visible area of an altimetry layer; (c) the subdivision of the region of interest into an indexed cells space; (d) Identification of the visible cells in the pyramid vision.

5 The Triangulation Construction Study Case

Applications that demand overall triangles appearance to be more regular, as terrain visualization, usually use Delaunay triangulation. The Figure 6a presents an extract of the altimetry archive related to Piraquara region (adjacent to Curitiba) at 1:25,000 scale. It corresponds to the MI-2842-4 chart, where contour lines had been generalized with a three meters tolerance by using Douglas-Peucker algorithm. After the generalization, this altimetry archive contains 2,478 points, against the 8,791 original ones. The Figure 6b presents the Delaunay triangulation containing 4,922 triangles.

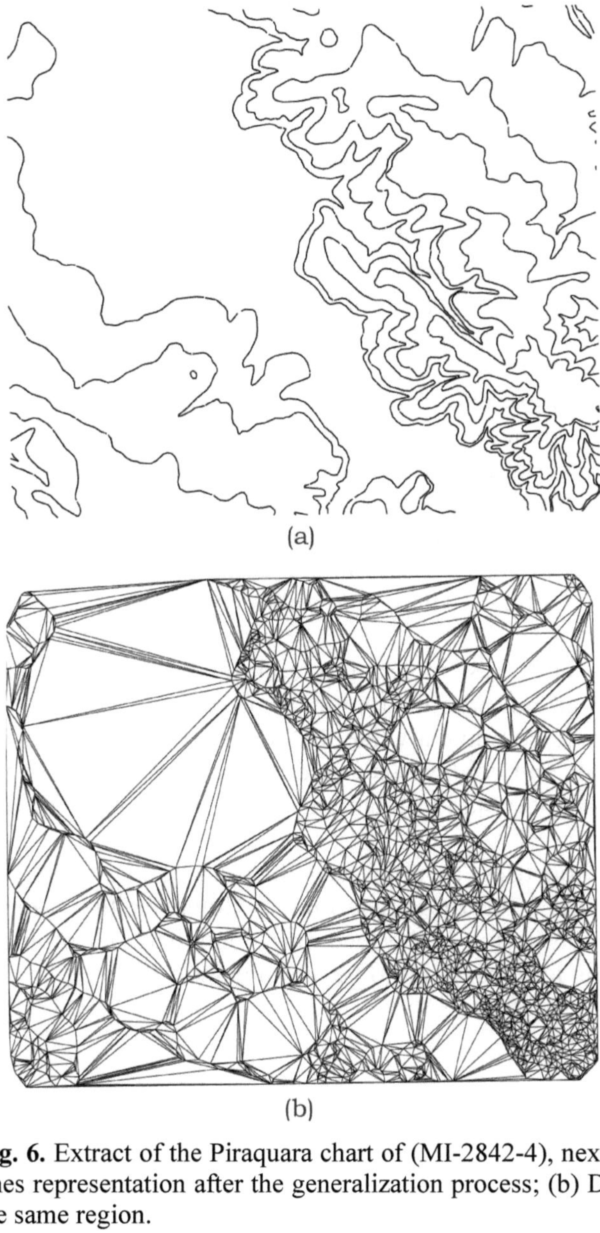

Fig. 6. Extract of the Piraquara chart of (MI-2842-4), next to Curitiba: (a) contour lines representation after the generalization process; (b) Delaunay triangulation of the same region.

The geometries, contained in each cartographic category, can be classified in sets of points, lines and areas. The point geometries constitute a trivial case of geometric intersection since it can only belong to a single cell. However, the lines and areas shapes are more complex cases. A line, for example, may have its initial and final vertex into distinct cells. Therefore, it is necessary to break the line into segments, each one of them belonging to a single cell [17].

Considering the region of the Figure 6a, for example, it could be divided into a 4x4 grid as illustrated in the Figure 7a. It should be noticed that this division implies in cutting several contour lines, forcing the curves segmentation in the cells frontier. Then, each cell could be triangulated using its own level curves, and the resulting mesh is presented in the Figure 7b. It can be observed the existence of holes at the grid corners, indicating that no triangulation covers those areas.

Despite the triangulation presents holes, it is known that the real terrain does not have them. Consequently, it is desirable to create some artifice that allows extending the triangulation for empty regions. An important property of Delaunay triangulation is that the sum, or combination, of a Delaunay triangulation set is a Delaunay triangulation too. So, the solution is to add four extra points to each cells point set, coincident to the delimited cells quadrilaterals vertices. These particular points are known as Steiner points [4].

The calculation of x and y Steiner points coordinates can be easily automated since the 4x4 grid divides the region of interest into equal areas. However, it is necessary to compute a value for Steiner point z coordinate in order to provide a soft and continuous aspect for terrain. It can be accomplished by using an interpolation technique such as the least square method.

The least square method uses, as inference function, points located in the neighborhood from the point to be determined. However, the volume of altimetry layer data can be too high, becoming computational prohibitive to locate those points for each point to be interpolated.

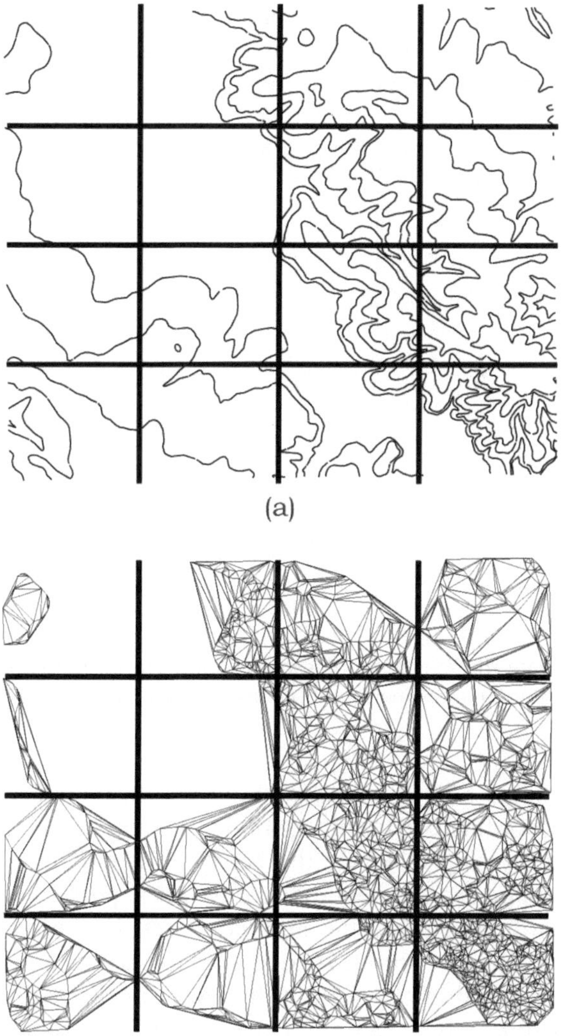

(a)

(b)

Fig. 7. Extract of the region of Piraquara: (a) the cartographic category of altimetry is divided according to a grating 4x4; (b) it is carried out the triangulation of the contour lines in each cell individually.

On the other hand, the indexing cell data structure can be used to perform an initial point set filtering. For each (x, y) Steiner point coordinate, it is possible to find out the adjacent cells. Subsequently, the search for neighboring points is performed only into the located cells, reducing the algorithm computational complexity. Figure 8 presents an altimetry layer scheme and its respective level curves. This layer was divided into a 4x4 grid, represented by the dashed lines. The black circle (Steiner points) represents a grid corner, which has quoted points into its neighboring cells. Therefore, the desired point z coordinate can be computed.

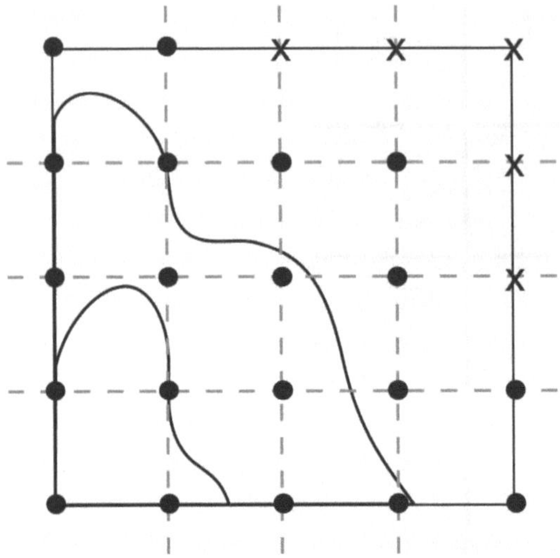

Fig. 8. The black points indicate the existence of sample points in its immediate neighboring cells, whereas the points designated with a x are surrounded by empty cells.

However, the cells that share Steiner points marked with x do not own any points to be used in z coordinate determination, that is, they are empty cells. In these cases, it is necessary to locate the empty cells neighboring patches, known as Steiner point second neighbors. Then, it is performed a new search for quoted coordinates using the new point set. In the worst case, this procedure may be repeated until it is necessary to execute a point search over the whole original altimetry layer. However, in the average case, this procedure tends to be better than a full search over the complete cartographic category.

After calculating the Steiner point's z coordinates, it is possible to compute a new triangulation for each cell. Thus, returning to the example from the Figure 7a, it is achievable to get the triangulation presented in Figure 9, containing 5,141 triangles. In this figure, the terrain mesh is subdivided according to a 4x4 grid, in which there are no emptiness regions, and the visual aspect looks continuous and soft allover.

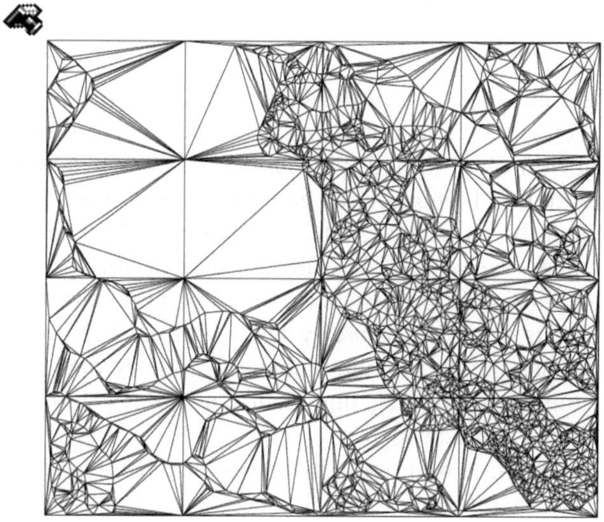

Fig. 9. Extract of the Piraquara chart altimetry layer, whose representative terrain mesh was subdivided into a 4x4 grid, where there are no empty areas.

Figure 10 represents the three-dimensional line of sight illustrated in Figure 9. The observer is located on the chart upper left corner, facing the lower right corner. In this figure it can be observed a soft and continuous surface, where Steiner points z coordinates were computed using the least square method. Moreover, by using a 4x4 grid, it could be reached an average of 16.40 fps (frames per second) in virtual flyovers using video boards without hardware acceleration.

Fig. 10. Three-dimensional line of sight of the Piraquara altimetry layer extract.

6 Results

The time spent by the graphical device to draw the scene into the monitor screen is inversely influenced by the quantity of geometrical primitives. The frustum culling algorithm, combined with Douglas-Peucker generalization, contributes for reducing the number of geometries necessary for the exhibition, and consequently, influences the system performance improvement. It was executed several tests in order to analyze the influence of the cells size on the three-dimensional visualization application performance. During these tests, just the altimetry layer was used. Nevertheless, it does not compromises the experiment because altimetry data represents about 76% of all the map data [2], that is, it contributes with great part of the geometric primitive of a scene.

During the carried out tests, it was used the complete MI-2842-4 chart (Piraquara), generalized with three meters tolerance, containing 137,841 triangles to be drawn. Two personal computers have been used. The first one had 256MB of RAM memory, 1GHz Pentium III processor, and the second equipped with 256MB of RAM memory and 233MHz Pentium I processor. The video boards used on the tests were: an S3 Virge of 1MB with PCI bus, an NVidia GeForce 2 of 64MB PCI bus and an NVidia GeForce 2 of 64MB AGP bus.

The S3 Virge video board does not implement any hardware accelerated instruction, whereas the GeForce 2 video boards do so. The video boards had been exchanged between the two computers, except the GeForce 2 AGP device which was not possible to install in the Pentium 233MHz computer because no AGB bus was available.

The region of interest was subdivided into 1x1, 2x2, 4x4, 8x8, 16x16, 32x32, 64x64, 128x128 grids, that is, 2^xx2^x grids, where x = 0,…, 7. Each one of these grids divides the chart into 1, 4, 8, 16, 64, 256, 1,024, 4,096, and 16,384 indexed cells. For each 3-tupla configuration <computer, video board, grid>, the same region of interest flyover was executed. This flyover performed random trajectories, passing through locations on edge and on the center of the chart. A complete description of these random trajectories can be found on Mello & Ferreira [18].

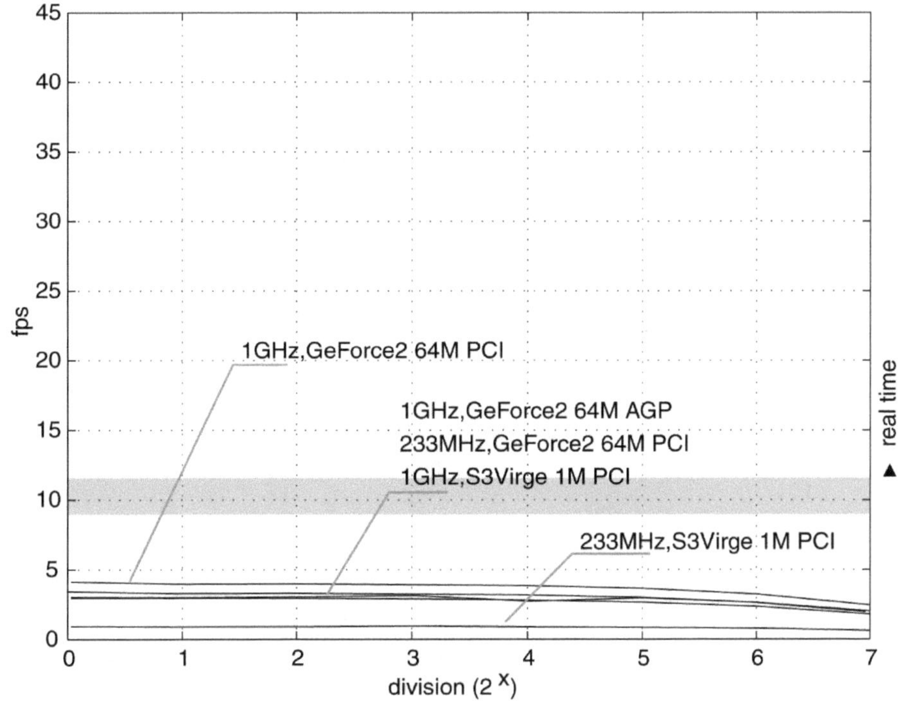

Fig. 11. Average values for video board refresh rate, without using cells culling, according to a 2^xx2^x grid subdivision.

The average values for video board refresh rate are illustrated on Figures 11 and 12. Figure 11 illustrates the results obtained by performing a flyover without the frustum culling algorithm, that is, all the geometries (triangles) had been drawn on each animation frame. On the other hand, Figure 12 represents the results associated to the invisible cells discard, that is, when only visible geometries were drawn.

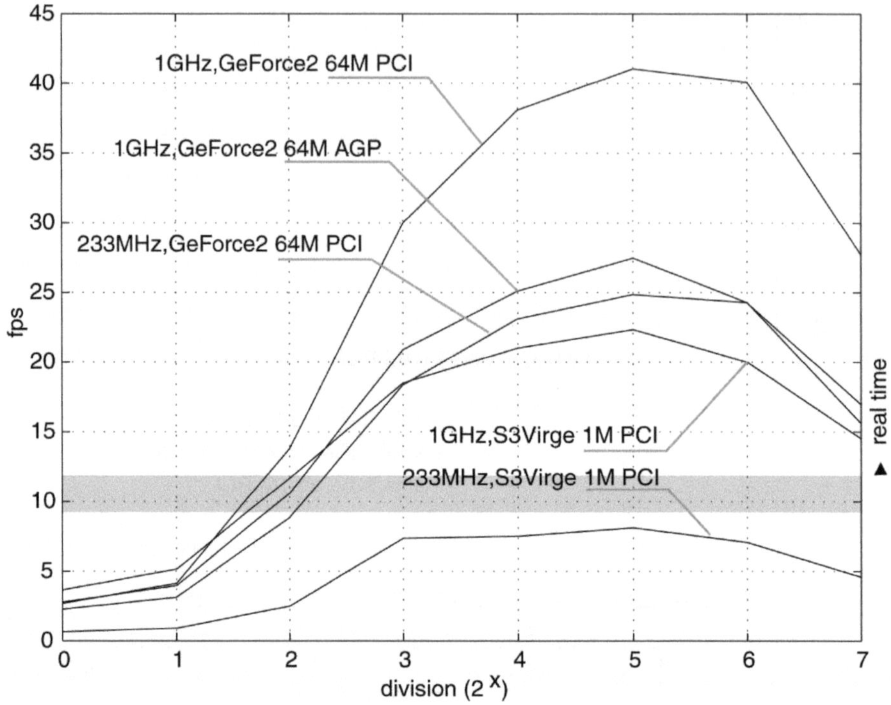

Fig. 12. Average values for video board refresh rate, using cells culling, according to a 2^xx2^x grid subdivision.

It can be observed that the use of a culling algorithm allows the configurations to upgrade from a non-real time performance zone to a real time one (except the configuration <233MHz Pentium, S3 Virge, *>), in accordance with the values defined by Lasseter [14].

The Figure 12 curves negative concavity also illustrates the relation between the increase of cells quantity and the video board refresh rate. Initially, the increase of grid divisions helps to enhance the video board drawing capacity. This occurs because the cost of the cells visibility test is compensated by the profit gotten with the invisible geometric primitive discarding in the scene. This situation is kept until an inflection point is reached, characterizing a state where the relation becomes the best. After that, a new situation is initiated, when it is observed video board frame rate depreciation. In this case, the computational cost to test all the discarding cells increases more quickly than the computational benefit of the discarding itself.

7 Conclusion

As previously mentioned, the relation between the terrain resolution and its associated data can easily exceed the video board processing capacity. In many cases, it becomes impossible to achieve a real time application. This problem becomes even more evident when trying to construct 3D terrain visualization applications based on data provided by national systematic mapping agencies. Also, it is a challenging task to implement 3D terrain visualization applications, based on real cartographical data, in a low processing availability environment.

Initially, the 3D visualization application was incapable to provide a video board refresh rate compatible to a real time system, in spite of the equipment used. Through the use of the clipping algorithm, it was possible to improve the application performance to higher values than the lower limits defined by classic animation literature. The results analysis suggests that it is possible to create a real time 3D terrain visualization application based on the current available cartographic data. The methodology for pre-processing cartographic data, proposed by Mello & Ferreira [17], proved to be useful to accomplish this goal.

References

[1] Allworth ST (1981) Introduction to Real Time Software Design. The MacMillan Press, New York.

[2] Anciães CLC (2003) Transformação entre redes Geodésicas: uso de coordenadas 3D, 3D com restrição, e 2D. *Dissertação de Mestrado*, IME, p.48.

[3] Astle D, Hawkins K, LaMothe A (2002) OpenGL Game Programming. Premier, California.

[4] Berg M, Kreveld M, Overmars M, Schwarzkopf O (2000) Computational Geometry: Algorithms and Applications New York, Springer, 2edn, c.9.

[5] Brasil (1994) Exército Brasileiro. IG 20-12: Sistema de Ciência e Tecnologia do Exército.

[6] Brasil (1998) Exército Brasileiro. C 7-1: Emprego da Infantaria.

[7] Copeland P (2004a) Power Sources For Future Infantry Suite, Defence Update, SSC Public Affairs Office Press Releases, n.1.

[8] Copeland P (2004b) Future Warrior: New Concepts for Uniform Systems, The Warrior Magazine, US Soldier System Center, Washington. November-December.

[9] Costa RSN (1992) Sistemas Baseados em Conhecimento em Engenharia de Processos. Dissertação de Mestrado, Instituto Militar de Engenharia, IME, Rio de Janeiro.

[10] Dwyer R (1989) A faster divide and conquer algorithm for constructing Delaunay triangulations Algorithmica 2: p.137-151.
[11] Flynn P (2004) Powering Up the High-Tech Soldier of the Future. InFocus, The National Academies, Washington, p.7-8. v.4, n.2.
[12] Fregosi AE, et al. (1980) Enfoque Clássico da Teoria de Controle. Campus, São Paulo.
[13] LaMothe A (1995) Black Art of 3D Game Programming. California, Waite Group, p.813-816.
[14] Lasseter J (1987) Principles of Traditional Animation Applied to 3D Computer Animation, Computer Graphics, p. 35-44. v.21, n.4.
[15] Martin J (1987) Design of Real Time Computer Systems, Prentice Hall.
[16] Mello FL, Ferreira LFC, Strauss E (2003) Um Sistema de Visualização Tridimensional do Teatro de Guerra. Revista Militar de Ciência e Tecnologia, p. 40-49. v.20, n.3.
[17] Mello FL, Ferreira LFC (2004) Visualização Tridimensional de Teatros de Guerra: o Pré-Processamento dos Dados. Revista Militar de Ciência e Tecnologia, p. 35-47. v.21, n.3.
[18] Mello FL, Ferreira LFC (2007) Análise de Desempenho de uma Metodologia de Tratamento de Dados Altimétricos Aplicada à Visualização Tridimensional de Terrenos. Revista Brasileira de Cartografia. (submitted)
[19] Shewchuk, JR (1996) Triangle: Engineering a 2D Quality Mesh Generator and Delaunay Triangulator, First Workshop on Applied Computational Geometry, ACM, Philadelphia, p.124-133.

A More Efficient Method for Map Overlay in Terralib

Vinícius Lopes Rodrigues[1], Marcus Vinícius Alvim Andrade[1], Gilberto Ribeiro de Queiroz[2], Mirella Antunes de Magalhães[1]

[1]Departamento de Informática
Universidade Federal de Viçosa
[2]Divisão de Processamento de Imagens
Instituto Nacional de Pesquisas Espaciais

1 Introduction

Geographical Information Systems (SIGs) are used to store, analyze and manipulate geographical data, that is, data representing objects and phenomena that have a geographical position associated to them which is essential to process and analyze them [2, 10, 13]. These systems involve problems from many areas such as computational geometry, computer graphics, database, software engineering, etc.

Recently, a group of Brazilian research institutes composed mainly by INPE (National Institute for Space Research), TECGRAF/PUC-RIO (Computer Graphics Group at the Catholic University in Rio de Janeiro) and PRODABEL (Information Technology Company for the City of Belo Horizonte) decided to develop an open source library, named TerraLib, whose main purpose is to enable quick development of custom-built GIS applications using spatial databases. In a general way, the Terralib aim is to enable the development of a new generation of GIS applications, based on the technological advances on spatial databases [2, 14].

The Terralib library is composed by many modules and, as any other GIS, it has a geometrical module that is responsible for giving support to geometric operations required by other GIS modules. Certainly, among these geometric operations, one of the most important is the overlay of two maps [1, 5, 11, 12]. As it is known, the map overlay is used in many situations when it is necessary to combine or to compare data stored in distinct information layers. For example, consider the following operation: "determine the deforestation areas inside indian reserves". To obtain these regions, it is necessary to combine two layers composed by many polygons: one layer defining the indian reserves and the other one defining the deforestation map. The overlay of these two layers gives a new map as shown in Figure 1.

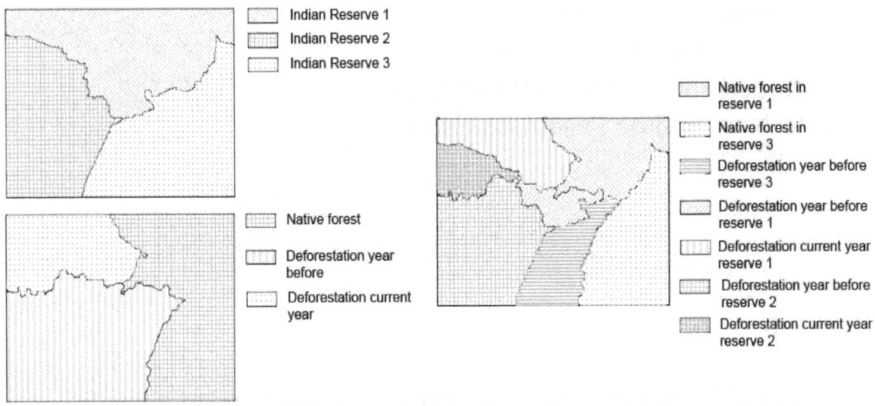

Fig. 1 *An application of map overlay operation.*

Since the overlay operation can be used many times and since it can involve maps with a lot of polygons then it is important to implement it as efficiently as possible. So, the main aim of this work is to describe the implementation of an alternative method to compute the overlay of two maps more efficiently than the original method in Terralib. In Section 4, some experimental results are presented to confirm the performance improvement.

2 The original overlay algorithm in Terralib

The original overlay algorithm included in Terralib, named *TeOverlay* operator, computes the overlay of two sets of polygons. This operator is used as the main component of a higher level operator called *TeGeoOpOverlayIntersection*. It is used to overlay two information layers and it deals not only with the geometries but also with the layers attributes.

This operator is general enough to be used in many situations. For example, the Terraview application [15] uses it to allow that the user combines two or more information layers.

Since the polygon overlay (*TeOverlay*) is a lower level operation, it is implemented in Terralib kernel. On the other hand, the *TeGeoOpOverlayIntersection* operator is implemented in the functions module.

More precisely, the Terralib does not provide a general method for overlaying two polygons. In fact, it provides a method to determine the union, the intersection or the difference between two polygons and this operation is implemented based on the algorithm proposed by Margalit and Knott [6]. A detailed description of this implementation can be found in Queiroz [8].

In the following, it will be described the *TeGeoOpOverlayIntersection* operator that has been used in Terraview to compute the overlay of two maps. It is important to note that the evaluation tests, described in the Section 4, were done comparing this operator with the proposed method.

Given two maps M_1 and M_2, suppose that M_1 is smaller (i.e. has less polygons) than M_2. The first step of this method is to determine the polygons of each map that have a chance of intersecting each other, that is, the polygons pairs whose bounding boxes have a non empty overlay.

The polygons candidates are determined connecting to the database manager (such as *MySQL, PostgreSQL, Oracle,* etc) used by the application and making SQL queries to recover them. More precisely, for each polygon P_i in M_1, a spatial SQL query is made to determine all polygons in M_2 whose bounding box overlaps the P_i bounding box.

Thus, in the next step, the operator *TeOverlay* is used to obtain the intersection points between the polygon P_i and the polygons candidates obtained in the last step. Using the intersection points obtained, the polygons' borders are fragmented and each fragment is classified related to the other polygon to identify if the fragment is inside, outside or on the border of it. This classification is done using a predicate to verify whether a point in the fragment is inside, outside or on the border of the polygon. Finally, using the fragments, the resulting polygons are built considering the desired operation (union, intersection or difference).

It is important to notice that the first step in this process involves database query operations and so, it can take a long time; mainly if the two maps have many polygons with many intersections, for example, overlaying two thematic maps of a same region.

Furthermore, a same polygon can be processed more than one time. For each polygon P_i used by *TeOverlay,* a same polygon can be identified as a candidate to intersect P_i.

3 An alternative (more efficient) map overlay method

To improve the efficiency of the map overlay method in Terralib, the basic idea was to adapt the original method redefining the process to obtain the intersection points between the polygons of the two maps. That is, the idea was to avoid the database queries made by the original method to obtain the polygons that have a chance of intersecting each other. To avoid these queries, the intersection points are obtained using an adaptation of the sweep line algorithm proposed by Bentley and Ottman [1, 4, 7, 9], described in Section 3.1.

The first step of the alternative method is to fetch the two maps from the database and to define two sets of segments: each set containing all segments of a map. So, the sweep line method is used to determine all the intersection points between the segments from each set. After that, the Terralib methods are called to build the regions corresponding to the map overlay – this operation uses the intersection points previously obtained. It is important to notice that the Terralib method was adapted to receive the intersection points as a parameter avoiding the database queries.

The region building process has to be executed for each pair of polygons that intersect each other and for an efficient identification, it is used a matrix where each position refers to a pair of polygons and it stores the list of intersection points between the pair of polygons. More precisely, given two maps having m_1 and m_2 polygons, it is defined a matrix with m_1 x m_2 positions and the matrix position (i,j) stores a list with all intersection points between the polygons i and j.

Thus, for each position (i,j) whose list is not empty, the adapted Terralib method is used to compute the overlay between the two polygons i and j.

This method receives the intersection points list as a parameter and the polygons' borders are fragmented and each fragment is classified related to the other polygon.

It is worth to say that, for each matrix position that has an empty list, it is necessary to verify if a polygon is entirely inside the other one and vice-versa. This test is done using a Terralib function that verifies if the polygons bounding boxes have a non empty overlapping – in this case, not only the bounding boxes' borders are considered but, most importantly, their interiors.

3.1 A method to determine the segment intersections

This method is based on the classical *sweep line method* proposed by Bentley and Ottman and consist of supposing that there is a vertical line *r* sweeping the plane, for example from left to right, and during this sweeping, some events are identified and treated – see Figure 2. Each event triggers a set of operations.

In this implementation, an event is represented by a point and the segment(s) associated to that point. For each event associated to a point *p*, there are three lists of segments (see Figure 2):

1. Left segments list: contains the segments that are on the left of *p*, that is, whose right extreme is *p*;
2. Right segments list: contains the segments that are on the right of *p*, that is, whose left extreme is *p*;
3. Intersections segments list: cointains the segments whose intersection point is *p*.

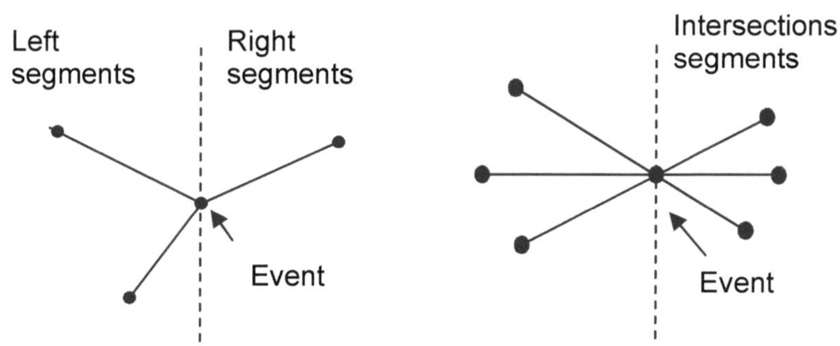

Fig. 2 *Events Examples.*

The events are stored in a list E, named *event scheduler,* that is sorted in non-decreasing order by the x values. This list is initialized with the extreme points of all segments.

The algorithm also uses a list L to store the segments that are intersected by the sweep line – these segments are called *active segments* and the list is sorted in non-decreasing order by the y values. The list L is dynamically updated when a event occurs: if an event corresponds to a left point, all the right segments are inserted in the list L; in right point events, all the left segments are inserted in the list L and in intersection point events, the intersection segments are reciprocally swapped – the first segment is swapped with the last one; the second with the second from the last, and so on.

When the list L is updated, the algorithm verifies if there is an intersection between the inserted or swapped segments in the list and their respective neighbor segment. Every intersection point obtained in this process is inserted in the list E and also in another list that will contain all the intersection points.

For sake of efficiency, the list of segments L is managed as a *Red-Black Tree* [3] to allow that the neighbors of a segment can be retrieved efficiently.

It is important to observe that, in the implementation of this algorithm, there are some special situations (degenerated cases) that have to be treated separately and, in general, they require a considerable effort.

A detailed description about the algorithm implementation can be found in [11, 12] where this same algorithm is used for an exact (round-off error free) manipulation of maps.

4 Results

The efficiency of the alternative method was evaluated by some experimental tests performed in a computer equipped with a AMD Athlon™ 1.31 GHz processor, 512 MB of RAM memory and running Microsoft Windows XP operation system.

The maps used in the tests are shown in Table 1 and Figure 3. The results obtained are shown in Table 2 and Figure 4.

As well as Terralib, the alternative method was implemented in Microsoft Visual C++ 6.0 and the data was stored in a Microsoft Access database, in Terralib format.

Table 1. Maps used in comparison tests

Number	Map	Number of regions	Number of segments
M_1	Cities MG	853	61237
M_2	Cities MG (translated$_1$)	853	61237
M_3	Temperature MG	20	4659
M_4	Soil MG	3067	858696
M_5	Cities Brazil	5560	571747
M_6	Cities Brazil (translated$_2$)	5560	571747

1 Translated 0,5 unit for the right in NoProjection projection of TerraView
2 Translated 1,2 unit for the right in NoProjection projection of TerraView

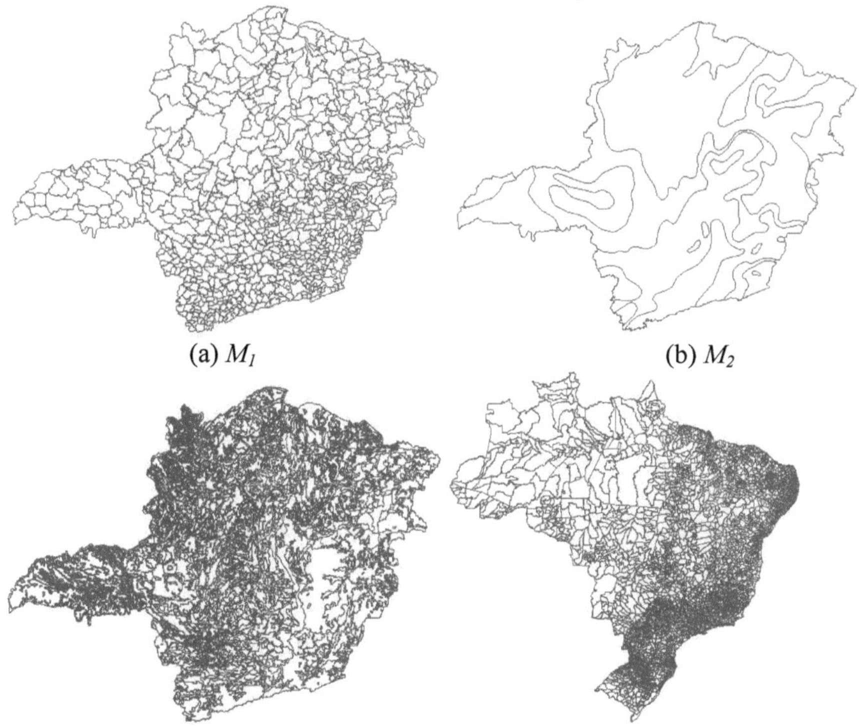

(a) M_1 (b) M_2

(c) M_4 d) M_5

Fig. 3 *Maps used in the evaluation tests*

Using these maps, the following test cases were performed (the resultant maps are shown in Figure 4):

Table 2 Comparison tests results.

	Number		Time (ms)	
Overlay	**Intersections**	**Output polygons**	**Terralib's**	**Alternative**
M_1 x M_2	15382	3885	184516	25037
M_1 x M_3	3678	1323	21591	18626
M_1 x M_4	39317	7437	433884	172007
M_5 x M_6	110232	25657	1554776	307913

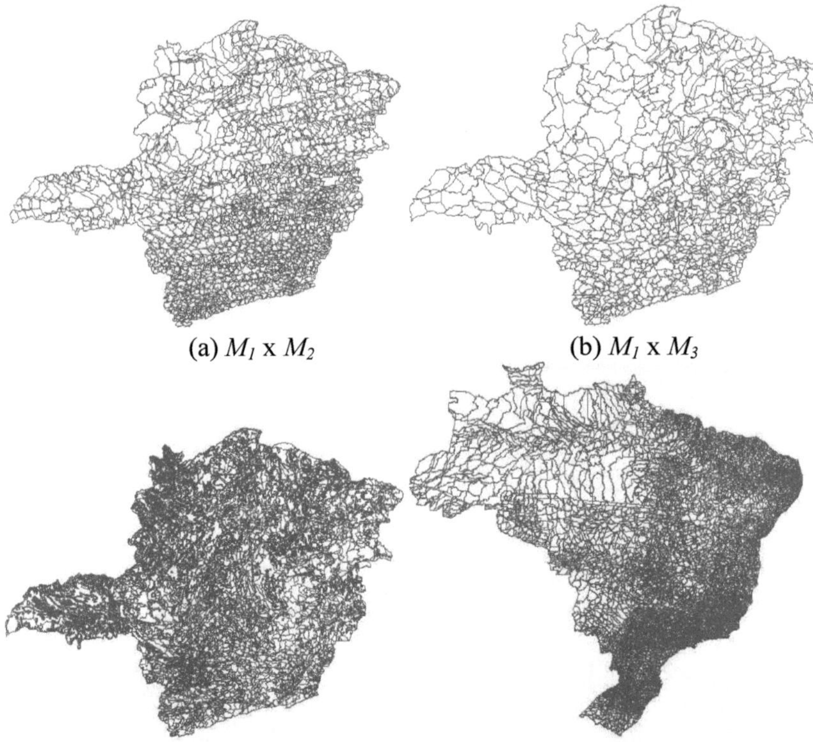

(a) M_1 x M_2 (b) M_1 x M_3

(c) M_1 x M_4 (d) M_5 x M_6

Fig. 4 *Obtained results*

As it can be realized, the execution times obtained by using the alternative method are considerably better than those obtained by the original algorithm. The differences become more evident as the size of the maps (number of segments and regions) grows up. For example, the overlay between the Cities of Brazil map and the same map translated took 26 minutes to be executed by the original algorithm while the alternative method obtained the same result in only 5 minutes.

5 Conclusions and Future Work

This work describes the implementation of an alternative method for map overlay in the Terralib library that uses the plane sweep paradigm to obtain the intersection points. As the tests showed, the proposed method is considerably more efficient than the original method. This improvement occurs mainly because the alternative method eliminates many spatial queries on the database. Besides that, the Terralib method obtains the intersection points between the polygons by testing an individual polygon against another, while the method presented in this work obtains all the intersection points processing all segments at once. Also, this approach avoids the treatment of a segment more than one time.

As a future work, the idea is to implement the overlay algorithm described by Kriegel [5] based on the same plane sweep paradigm. The expectation is that the results will be better than those shown in this work.

6 Special Thanks

This work was partially supported by CNPq (National Counsel of Technological and Scientific Development) and FAPEMIG (Research Foundation of the State of Minas Gerais).

References

[1] Berg M, Kreveld M, Van Overmars M, Schwarzkopf O (2000) Computational Geometry, Algorithms and Applications. Springer Verlag.
[2] Câmara G. et al (2000) Terralib: Technology In Support of GIS Innovation. II Workshop Brasileiro de Geoinformática, Geoinfo, Caxambu.
[3] Cormem TH, Leiserson CE, Rivest RL (1990) Introduction to Algorithms. The MIT Press.
[4] Daltio J, Andrade MVA, Queiroz GR (2004) Tratamento de Erros de Arredondamento na Biblioteca TerraLib. VI Simpósio Brasileiro de Geoinformática, Campos do Jordão, SP.

[5] Kriegel HP, Brinkhoff T,.Schneider R (1991) An Efficient Map Overlay Algorithm Based on Spatial Access Methods and Computational Geometry, Workshop Proceedings, Capri.
[6] Margalit A, Knott GD (1989) An algorithm for computing the union, intersection or difference of two polygons, Computers & Graphics, v. 13, n. 2, p. 167-183.
[7] Preparata F, Shamos M (1989) Computational Geometry: An Introduction. Springer-Verlag.
[8] Queiroz GR (2003) Algoritmos Geométricos para Bancos de Dados Geográficos: da Teoria à Prática na Terralib. Master thesis, INPE.
[9] Rezende PJ, Stolfi J (1994) Fundamentos de Geometria Computacional. IX Escola de Computação.
[10] Rigaux P, Scholl M, Voisard A. (2002) Spatial Databases with Applications to GIS. Morgan Kaufmann.
[11] Rodrigues VL, Andrade MVA, Queiroz GR, Daltio J (2006) Algoritmos exatos para interseção de segmentos e para sobreposição de mapas incorporados à biblioteca TerraLib. 32a Conferencia Latinoamericana de Informática, Santiago, Chile.
[12] Rodrigues VL, Cavalier AS, Andrade MVA, Queiroz GR (2005) Algoritmos exatos para tratamento de mapas na Terralib. VII Simpósio Brasileiro de Geoinformática, Campos do Jordão, SP.
[13] Steinberg, SJ, Steinberg, SL (2006) GIS: Geographic Information Systems for Social Sciences: Investigating Space and Place. SAGE.
[14] TerraLib home page (2006) http://www.terralib.org.
[15] TerraView home page (2006) http://www.dpi.inpe.br/terraview

Exploiting Type and Space in a Main Memory Query Engine

Thomas Schwarz, Matthias Grossmann, Daniela Nicklas, Bernhard Mitschang

Institut für Parallele und Verteilte Systeme
Universität Stuttgart

1 Introduction

In the upcoming areas of location-based services and ubiquitous computing new data-intensive applications emerge, which support their users by providing the right information at the right place, i.e., providing on demand what fits best to the user's current situation. Usually, the user's position and the application he is currently using determine the relevant information, so most information requests issued by the application contain spatial predicates and predicates restricting the type of the data. In this paper, we present a dedicated main memory query engine that is tailored to this environment and that supports application-specific processing capabilities. In particular, we analyze which index structures are best suited to maximize its performance.

The idea for this query engine emerged from the experiences with a data and service provisioning platform for context-aware applications. Data providers manage spatially referenced data, e.g., rooms, facilities, and sensors in a building, or the map data of a city. There, several data management systems that are specialized to the characteristics of the managed data (i.e., update rate and selection usage) [9] have been developed. In order to combine the data of multiple providers an integration middleware [24] has been developed. It achieves a tight semantic integration of the data instances using an extensible integration schema [16]. A plug-in concept allows to employ domain-specific functionality in the middleware like detecting duplicates, merging multiple representations, or aggregating and generalizing (map) data. The platform is used by various location-based applications like a city guide (a tourist application) or a digitally assisted scavenger hunt (a multiplayer mixed reality game) [17]. According to our experience, applications get by with simple selection queries.

Our query engine is also of interest to others. It can be directly integrated into implementations of the OGC Catalogue Services standard [18] or within the FGDC clearinghouse [15], which both offer a discovery mechanism for digital geospatial data. Similarly, implementations of geographic information systems may profit from our query engine. Furthermore, grid metadata catalog services [26] or discovery services in a service-oriented architecture [2] can apply our approach in order to optimize their engines that select different types of resources or services based on given restrictions.

1.1 Contribution

In this paper, we describe the design and implementation of a main memory query engine employing an index structure that leverages spatial dimension and type dimension, such that location-conscious queries are most efficiently supported. The focus is not on indexing and index structures, but on configuring the query engine's internal data structures to exhibit the best possible index organization. In order to do this, we evaluate two different approaches to organize an index structure that combines a spatial dimension and a type dimension. We detail on three different variants to map the type information (type IDs) to values in the type dimension. This has a substantial impact on the performance of the query engine, but has not been considered previously. We also point out how to determine the best range for the mapped values.

Many components in a large-scale information system may profit from the proposed query engine. Therefore, we describe a solution architecture for such an information system and introduce four different usage scenarios for four of its components, each having different characteristics. In order to achieve a sub-second response time of the overall system (including network latencies, (de)serialization and other processing overhead) the individual query engines have to process a typical query returning about 1,000 objects in 10 milliseconds, as a rough estimate. Therefore, we emphasize a main memory approach in order to achieve fast response times and allow for an easy deployment.

We run a substantial number of experiments and assess the suitability of the various techniques specifically for each scenario. Compared to an approach using separate indexes on type and space we can increase the performance up to almost an order of magnitude in certain cases. Our goal is to enable the reader to apply our insights profitably to his problem at hand.

The remainder of this paper is structured as follows. In Section 2 we introduce the typical data managed by our query engine and the typical queries issued. In Section 3 we characterize its usage scenarios. We describe the approaches to organize the index structures used by our query engine in Section 4. In Section 5 we describe the conducted experiments and analyze their results. We give an overview on the related work in Section 6. Finally, we conclude the paper and indicate future work in Section 7.

2 Data and Queries

Typically, applications in the domain of location-based or context-aware applications operate on object-structured data, see Figure 1. In the GIS world, objects are also called "features". The schema consists of a collection of types. An object is associated to a type which determines the name and data types of the attributes that an object of this type may use to store information. Types are structured in a is-a-hierarchy, see Figure 1 for a typical example.

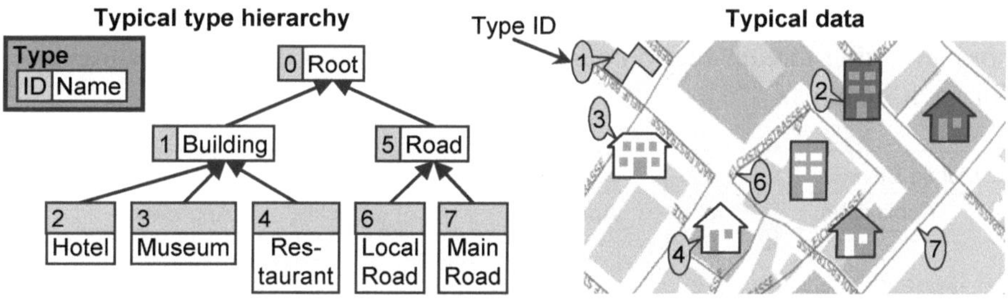

Fig. 1. Simplified excerpt of a typical type hierarchy (schema) and typical data

We assign a unique number called type ID to each type. Using an optimal assignment (termed linearization in [14]) we are able to determine for each type a continuous interval that contains exactly the type's own ID and the IDs of the subtypes of this type. This works always for type hierarchies with single inheritance [14].

We assume that in the targeted application domains every object has a position or an extent so that already the root type of the hierarchy comprises a generic geometry attribute that we exploit for indexing purposes. Examples for such schemas are the TIGER/Line data model [27], augmented world models like the one used in [16], or the upcoming standard for city models, CityGML [11]. Objects have linestrings and polygons as geometries, which all can be approximated by bounding boxes from an indexing point of view. Typical data sets comprise various kinds of roads (local road, main road, highway, ...), buildings, points-of-interest (museum, church, viewpoint, ...), and so on, see Figure 1 for some ideas.

Expressed in natural language, typical queries are "Give me all roads (no matter what kind) in the given rectangle", "Give me all major roads in the corridor between my current and my target position", or "Give me all French restaurants within 1 mile". All these queries have in common that they have a spatial predicate restricting the position of the result objects and a type predicate restricting the type of the result objects. Usually, the query addresses also all subtypes of the sought type. Therefore, we strive for exploiting this commonality by supporting such queries with a tailored index approach.

3 Usage Scenarios

An information system can employ the proposed query engine in various ways and places. We focus on location-based and context-aware systems that integrate data dynamically from many data providers ranging from web sites over digital libraries and geo-information systems to sensors and other stream-based sources. Figure 2 shows a typical architecture for such systems.

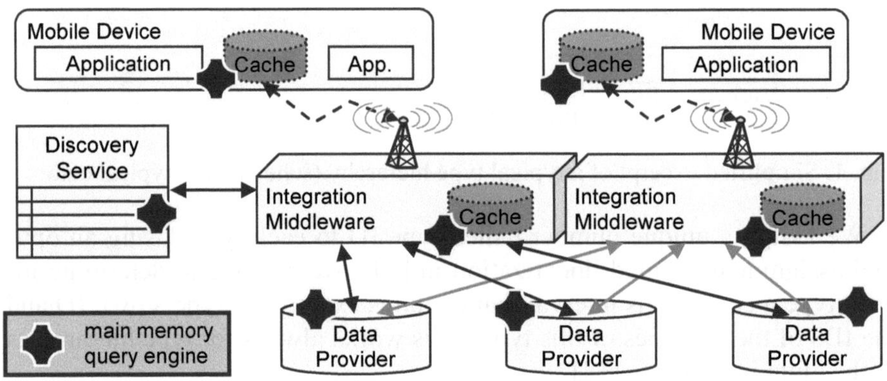

Fig. 2. Architecture of a location-conscious data provisioning system integrating data from various providers

The processing model is as follows. An application on a mobile device issues a query for data relating to the user's vicinity. The query is first processed by the local query engine. A query for the missing data is issued to the integration middleware. There, the query engine of the middleware processes the query. For retrieving the missing data the integration middleware determines the relevant providers using the discovery service, which itself runs a query engine. The integration middleware requests the data from these data providers. They evaluate location-based queries using a query engine as well. When integrating their results a query engine supports the integration middleware in evaluating additional predicates and performing location-based data merging. Finally, the integrated result is sent to the application. On all levels query processing can be considerably enhanced by means of data caches (cf. Figure 2).

As mentioned before, those query engines basically consist of a specific index structure that supports predicate-based queries that predominantly consist of location and type predicates. Hence, the efficiency of the query engine is mostly determined by the performance of the supporting index structure. Obviously, all of the above four types of components (mobile device, integration middleware, discovery service, and data provider) running such query engines do benefit from the index structures we investigate in this paper.

However, each component manages a different piece of the data, has different typical queries, and updates or exchanges the data in a different way and at a different frequency. Hence, we analyze the experiments in Section 5 individually for each component. Table 1 summarizes and quantifies these characteristics, which have been derived from the experiences with our service provisioning platform for context-aware applications [24].

Table 1. Selectivity factors (SF) and update rates (number of updated objects per number of queries) of the usage scenarios

Usage Scenario	Spatial SF	Type SF	Update rate
Data Provider	1% - 20%	20% - 100%	0.01
Discovery Service	1% - 5%	1% - 20%	0.1
Integration Middleware	10% - 50%	1% - 20%	10
Mobile Device	10% - 50%	10% - 100%	100

The term selectivity factor (SF) refers to the ratio of objects qualifying for the result set to the total number of objects (the data set size). If a predicate has a low SF then only few objects qualify for the result set, and vice versa. The Spatial SF refers to the ratio of the area of the query window to the area of the data set's universe, which is given by the convex hull around the geometries of all objects in a data set. Update rate counts the number of objects that are updated between two consecutive queries.

E.g., an update rate of 0.1 means that only one object is changed during a period where ten queries are processed.

4 Index Structures

In this section we give details about the different approaches that we investigated to build an index structure that combines the spatial dimension and the type dimension. This combination is a natural consequence from the observation in Section 2 that the majority of queries involves selection predicates on at least these two dimensions. For each approach we explain how the query engine processes a query step by step.

We refer to the spatial dimension as a single dimension, although it actually involves two dimensional coordinates. Also, we will abstract from the details of particular spatial index structures (e.g., R*-Tree, Grid-File, or MX-CIF Quadtree, see [8] for a survey) because the underlying spatial index structure can be easily exchanged without significantly shifting the relative performance of the presented approaches. We focus on how to combine existing well-known index structures in new ways to best solve the problem at hand.

The following approaches are designed to work in main memory, which is a requirement to achieve reasonable response times. Partitioning techniques can be applied to split the data into chunks that a single system can maintain in main memory. Furthermore, this allows us to flexibly deploy our query engine to any component in the entire system with very little administrative overhead in contrast to deploying a full fledged database system.

All approaches use hash data structures that map a type's name to its ID, and this ID to a list of the IDs of all sub or super types in constant time. The size and contents of these lookup tables depends only on the type hierarchy so that they are small in size compared to the entire data set and they are not affected by updates.

4.1 Separate Indexes (SEP)

The Separate Indexes approach maintains two distinct data structures, see Figure 3. The first data structure uses a spatial index to organize the objects solely by their geometry. The second data structure uses an array containing for each type a separate list of the corresponding objects. Objects are inserted into both data structures, which have to be in sync at all times.

For answering a query a cost-based optimizer assesses the selectivity of the spatial predicate and the type predicate. Then, the more selective predicate (lower selectivity factor) is used to generate a list of candidates using the corresponding data structure. Finally, these candidates are filtered using the remaining predicate.

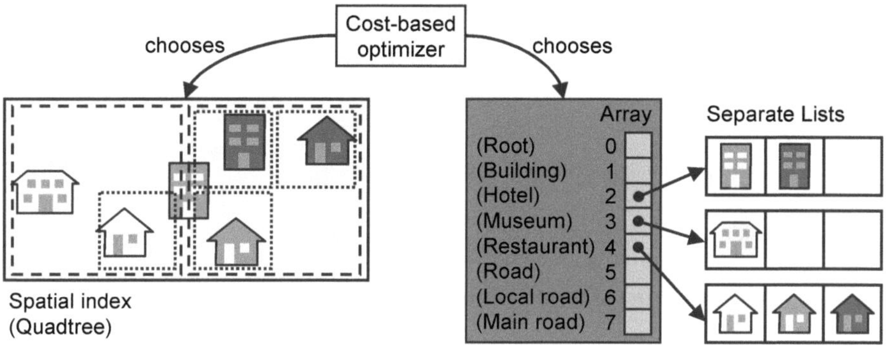

Fig. 3 . Data structures in the Separate Indexes (SEP) approach

This approach gives us the bottom line of the least achievable performance as it uses only standard database technology without any problem-specific improvements. This approach tends to be slow because it exploits an access path for at most one dimension and filters all candidates along the other dimension.

4.2 Real 3D Index (R3D)

The Real 3D Index approach is especially tailored to the typical queries introduced in Section 2. It maintains a single spatial index that involves three orthogonal dimensions, see Figure 4. Two dimensions are used to store the bounding boxes of the objects' geometries. The third dimension is used to store the objects' type ID. Each object is inserted into this index only once.

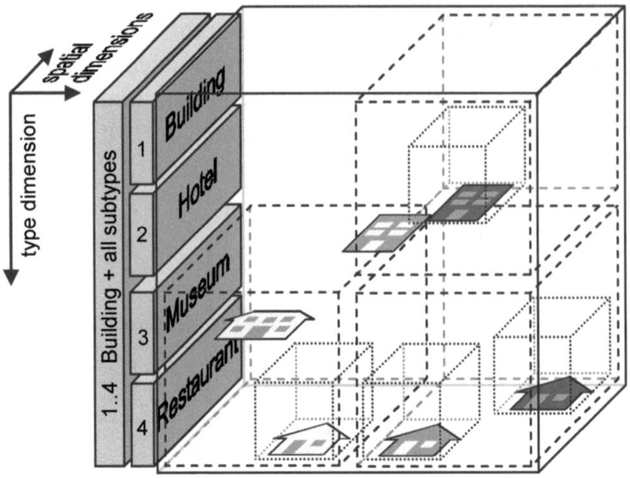

Fig. 4. Data structures in the Real 3D Index (R3D) approach

In this approach a query involving a spatial predicate and a type predicate can be expressed as a three dimensional bounding box where the range in the type dimension comprises the ID of the sought type and the IDs of all of its subtypes. Thus, each query can be translated into a single bounding box that is used for a single traversal of the Real 3D Index.

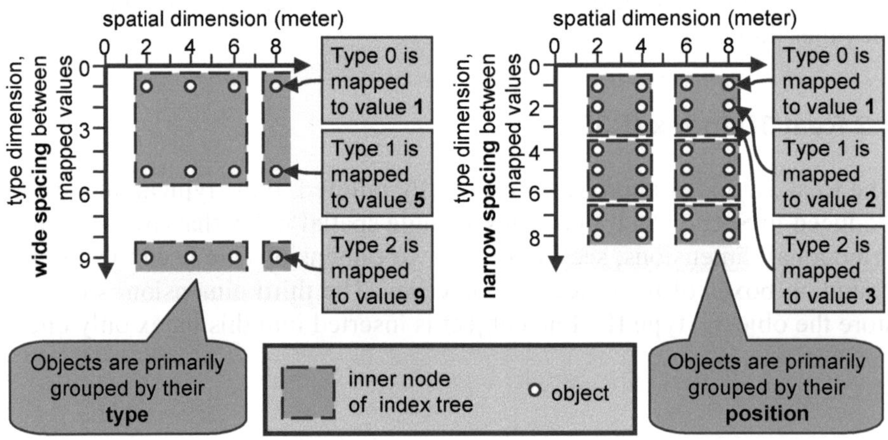

Fig. 5. Effects of the spacing between mapped type values in the type dimension on the clustering of objects in the inner index nodes

The mapping of types to values in the type dimension is the critical aspect in this approach. As shown in Figure 5 the space between the mapped values of two adjacent types influences the clustering of objects and child nodes in the inner nodes of the index tree. If there is a large gap between the mapped values of two types (wide spacing, left part of Figure 5), then objects are grouped by their type value rather than by their position in the spatial dimension. In the example objects with three different positions and only two different types are grouped in the same inner index node. If the mapped values of two types are close to each other (narrow spacing, right part of Figure 5), then it is vice versa. Inner index nodes store objects with only two different positions but three different types. This is due to the fact, that most indexing methods try to keep the bounding boxes of the inner nodes as squarish as possible. As we will see in Section 5, the spacing of the values in the type dimension has a huge impact on the performance of the index. It determines if whole branches can be pruned away when traversing the index tree. With wide spacing, subtrees with the wrong type can be skipped quite early. With narrow spacing, the same goes with subtrees having too distant positions.

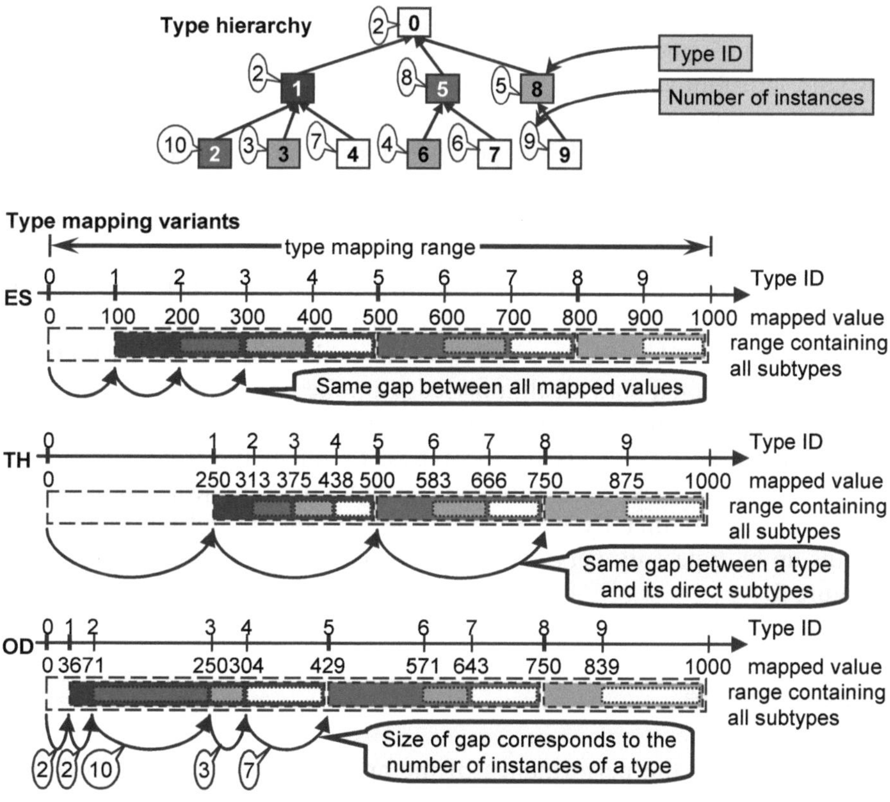

Fig. 6. Computing the mapped type values for the different type mapping variants

We have investigated three different kinds of spacing, which are visualized in Figure 6. In all variants, the mapped values are scaled to span a predetermined type mapping range, which is about as large as the average distance between neighboring objects multiplied by the number of types in the type hierarchy, see Section 5.2 for more details. The dashed boxes indicate the range containing the mapped values of a type and all of its (transitive) subtypes.

- Bottom-up equal spread (ES): Each type's value in the type dimension equals the type's ID multiplied by a constant scaling factor. This is the simplest variant which disregards the type hierarchy to a large extent. It is a close relative to the approach pursued in [14]. In the example in Figure 6 we have 10 types and the type mapping range is 1000 units, so that the distance between two mapped values is 100 units.
- Top-down biased by type hierarchy (TH): On each level of the type hierarchy the available mapping range is evenly distributed among the current type and its subtypes. In Figure 6, type 0 has three direct subtypes (1, 5, and 8), so that the available mapping range of 1000 units is split into four segments. Type 5, in turn, has to split its range (500 to 749) into three segments because it has two subtypes. This variant preferredly groups objects having the same supertype in the inner tree nodes.
- Top-down biased by object distribution (OD): On each level of the type hierarchy the available mapping range is distributed among the current type and its subtypes based on the number of instances each type has. In Figure 6, we have a total of 56 object instances. Two object instances have type 0, so that the gap to type 1 is units. Note that this method requires additional statistical knowledge on the object distribution. This variant groups objects in the inner tree nodes according to the actual distribution of the objects on the types. Frequent types get their own subtree already close to the root of the index tree. Objects having rare types are predominantly grouped by their geometry and the index tree splits up by type only very close to the leaves of the index tree, see also Figure 5.

5 Experiments

In order to assess the performance of the approaches we implemented all of them in Java. We used the MX-CIF quadtree [23] implementation of the JTS Topology Suite [31] as our spatial index structure and adapted it to cope with more than two dimensions. We added some optimizations so that in the end it was faster than the XXL library's [30] R*-Tree implementation for both inserting and querying objects. However, we point out that the actually used spatial index method has only a marginal influence on the relative performance of the different approaches.

We used a dual processor Dell workstation having 2GHz Intel Xeon processors and 2 gigabytes of RAM, half of which was assigned to the Java virtual machine. All experiments were run on a single processor while the other one was idle to minimize disturbances caused by the operating system. In order to get a reasonable precision when measuring submillisecond response times we used the high resolution timer package [22].

We conducted the experiments using a subset of the TIGER/Line 2003 data sets [27]. In particular, we ran queries against the data sets of nine counties in California, see Table 2 for their characteristics. We extracted the linestring and polygon based features leading to data sets comprising between 12k and 200k objects.

Table 2. Data sets used in the experiments

Abbreviation	County	Size	Universe (in km)		Number of objects
			Width	Height	
#1	Yuba	Small	33.2	27.6	11923
#2	Glenn	Small	42.6	70.8	16839
#3	San Francisco	Small	15.0	78.4	22666
#4	Alameda	Medium	37.6	49.7	46285
#5	Santa Clara	Medium	42.1	42.8	53727
#6	Sacramento	Medium	49.2	45.5	71743
#7	Riverside	Large	59.0	26.5	151489
#8	Kern	Large	98.6	17.0	175082
#9	San Diego	Large	90.6	114.8	203122

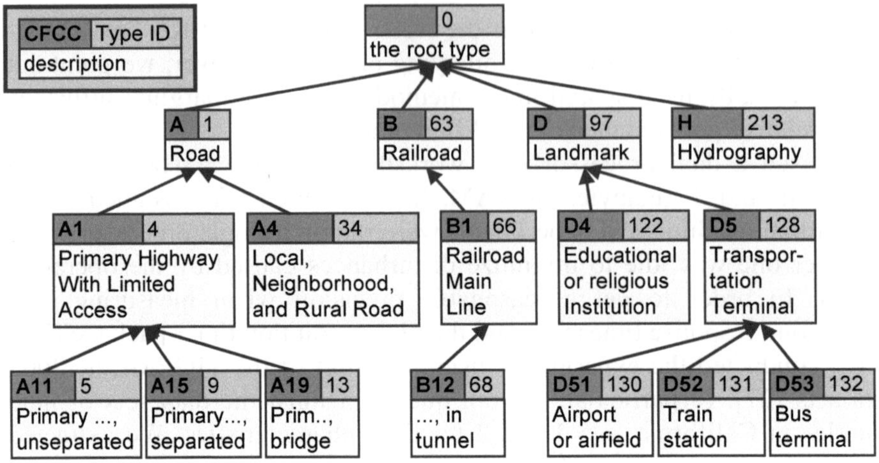

Fig. 7. Excerpt of the type hierarchy based on the CFCC feature codes used in the TIGER/Line data sets

The data sets contain data about roads, railroads, other ground transportation, landmarks, hydrography, property boundaries, etc. We interpreted the CFCC feature codes [28] as type names and built up a type hierarchy with four levels, see Figure 7. The single lettered CFCC codes constitute the direct children of the hierarchy's root node. The codes having two letters make up their children, and so on. In total, the type hierarchy consists of 258 nodes.

5.1 Computing the total average query and update response time (TAQURT)

As a system using our query engine cannot switch between different indexing approaches on the fly, we do not compare individual measurements. Instead, we compute a weighted average response time for each combination of indexing approach, usage scenario, and data set. For this, we picked a representative subset of all types and a set of ten differently sized query areas. For each combination of type (fixed type SF) and query area (fixed spatial SF) we measured the response time, leading to a grid of measurements. We interpret the measurements as support points for a piecewise linear surface function having the spatial SF and the type SF as the independent variables. For each approach, usage scenario, and data set we compute the weighted average query response time by computing the weighted integral of the surface function along the axes of the independent variables. The usage scenarios provide the parameters describing typical workloads. Their minimum and maximum SFs (see Table 1) define the integration limits. The surface function is weighted by the reciprocal of the multiplied selectivity factors. The rationale behind this is that small queries retrieving only few objects are more frequent than large queries. Calculating the weighted average response time by integrating the piecewise linear surface function has the benefit of allowing to arbitrarily set the minimum and maximum selectivity factors. Furthermore, the integral allows to calculate a more meaningful average value that takes each measurement's SF into account.

Finally, the usage scenario also determines the frequency of updates. The total average query and update response time (TAQURT) is the sum of the average insertion cost per object weighted by the update rate and the average query response time:

Thus, we get a TAQURT for each combination of indexing approach, usage scenario, and data set. By aggregating the query and update performance into a single figure we can evaluate the approaches from a "total cost of ownership" perspective and by concentrating on the four usage scenarios we keep the experiments clear.

5.2 Comparing the type mapping variants of the Real 3D Index

In this section, we compare the three mapping variants equal spread (ES), type hierarchy (TH), and object distribution (OD) introduced in Section 4.2 in order to determine the best one. The mapping variants differ in the spacing in the type dimension between the mapped values of two adjacent type IDs. Taking the type IDs themselves as the mapped values in the type dimension (1:1), as proposed in [14], is a very bad idea that leads to an average performance loss between 37% and 418%, see Figure 9.

We vary the range of the mapped type values in five steps, denoted A, B, C, D, and E, see Table 3. The sizes of the selected mapping ranges are around the order of magnitude of the anticipated optimal mapping range, which is approximated by calculating the average distance of the objects of one type along one of the spatial axes (4200 meter in our data sets) and multiplying it by the number of types in the hierarchy (258). This way objects are clustered equally along the spatial dimension and the type dimension in the inner nodes of the index tree. The selected mapping ranges are quite representative as in most cases one of the medium ranges shows the best performance, see Figure 8.

Table 3. Type mapping ranges

Mapping range	A	B	C	D	E
Total range (in kilometer)	15	150	1500	15000	60000

We compute the columns displayed in Figure 8 as follows. We group the measurements by usage scenario and data set. For each combination of mapping variant and mapping range we compute its relative response time as the ratio of its TAQURT to the minimal TAQURT in each group. Then, we average the relative response time across all data sets.

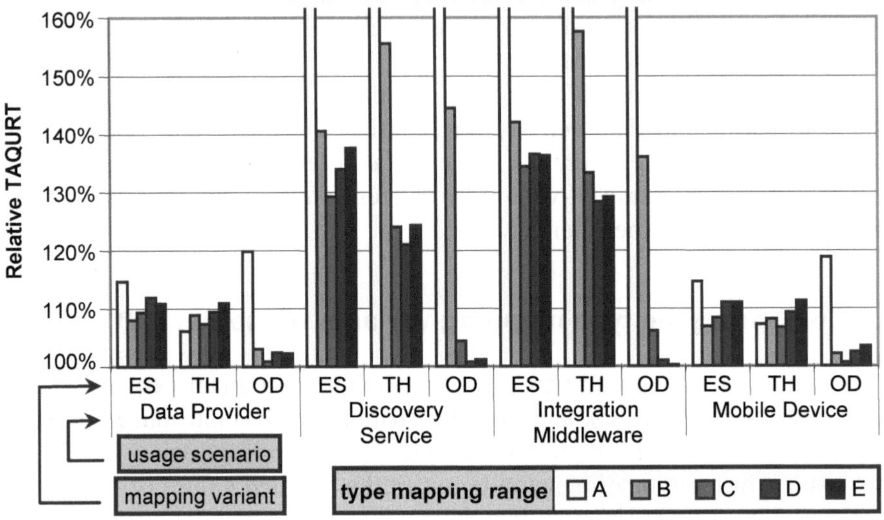

Fig. 8. Relative total query and update performance of the Real 3D Index mapping variants for each scenario

The first observation is that the data provider and mobile device usage scenarios are less sensitive to the mapping range than the other two scenarios. This is due to them having a high type SF. The main insight of this figure is that it suffices to get close to (within an order of magnitude) the best possible mapping range to get reasonable performance (5% worse than best possible). However, if you are far off the mark (mapping range A) then the performance degrades considerably. Unfortunately, the best mapping range differs depending on the usage scenario and mapping variant. Averaged across all scenarios, the ES variant achieves the best performance with mapping range C. The TH and OD variants work best with mapping range D.

Figure 9 displays a subset of the results shown in Figure 8. For each mapping variant in each usage scenario only the column corresponding to the mapping range that yields the fastest relative TAQURT is displayed. Additionally, they are compared to the 1:1 mapping variant and to the Separate Indexes approach (SEP).

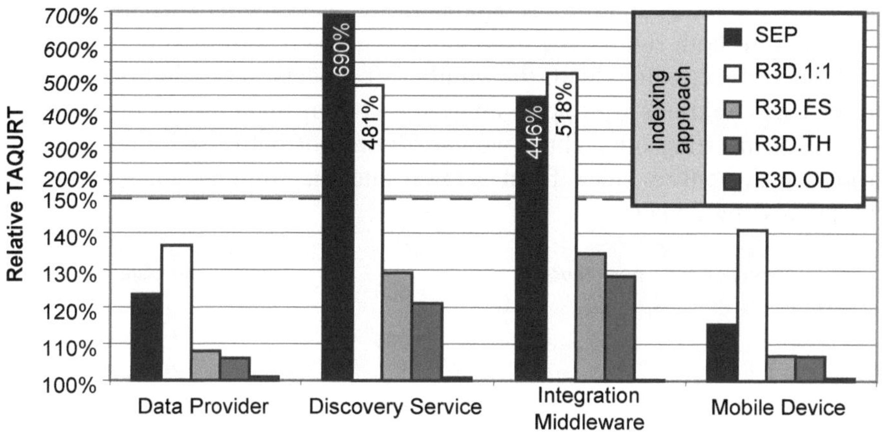

Fig. 9. Comparing the best mapping ranges of Figure 8 with the 1:1 mapping variant and the SEP indexing approach for various usage scenarios

Figure 9 clearly shows that the OD variant delivers the best performance. The data provider and mobile device usage scenarios both have a high type SF and the performance differences are less pronounced there. The OD variant leads there by about 7%. In the other two scenarios the lead is at least 20%. However, the OD variant has the disadvantage that it needs advance knowledge about the distribution of the objects on the types of the type hierarchy.

The best alternative to the OD variant is the TH variant, which is always between 1% and 8% better than the ES variant. The 1:1 variant is far behind and in most cases it is even worse than the SEP approach. This highlights the importance of choosing an adequate type mapping range. Doing so gives the R3D approach a comfortable lead over the SEP approach in the usage scenarios having a high type SF. When the type SF is low the R3D approach really outperforms the SEP approach by almost an order of magnitude. In the discovery service usage scenario, where spatial SF and type SF are both low, even the 1:1 mapping variant of the R3D approach is faster than the SEP approach.

5.3 Index construction

In this section we analyze the costs involved with maintaining each index structure. The costs are divided into the average time needed to insert an object into the index and the average memory occupied by each object (see Figure 10). In both figures the total values can be determined by multiplying the per object values with the number of objects in the data set. In our further considerations we approximate the cost of updating an object with the cost of inserting one. Thus, we can figure out both the initial cost for setting up the entire index from scratch and the running costs involved with processing updates.

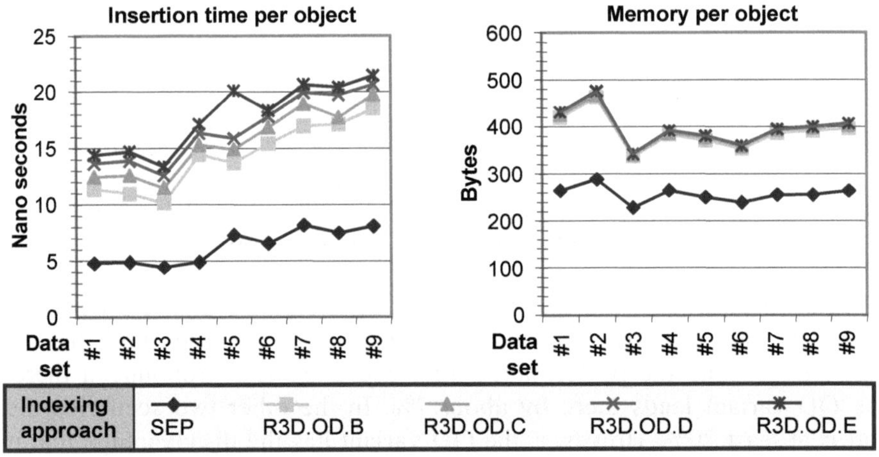

Fig. 10. Average time for inserting a single object into the index, and average memory occupied by a single object (including geometry, bbox, type ID, and indexing overhead)

Figure 10 shows for the Separate Indexes approach (SEP) that the time per object increases only slightly with an increasing data set size, so that the approach scales quite well. The Real 3D Index (R3D) is 2.4 to 3.2 times slower than the SEP approach for small data sets. For large data sets it is only 2.2 to 2.6 times slower than the SEP approach. As shown in Figure 10, the type mapping range (B to E) has a significant impact on the index creation time. The larger the type mapping range is, the more time is needed to build the index. The other type mapping variants (ES and TH) not shown in Figure 10 behave similarly.

The average amount of memory (including indexing overhead) occupied by a single object is about the same for all data sets, see Figure 10. Large data sets do not lead to worse memory utilization. The least memory is occupied by the SEP approach. All R3D variants have about the same memory footprint. They occupy about 50% more memory than the SEP approach.

To put it in a nutshell, the SEP approach clearly outperforms the R3D approach in terms of only index maintenance costs. It uses less memory and is faster in building up the index.

6 Related Work

As indicated in the introduction, OGC Catalogue Services [18] (OGC CS) and Grid Metadata Catalog Services [26] (MCS) can benefit from the proposed query engine. Currently, vendors like ESRI, Galdos, or Ionic offer geodata catalog services complying with OGC CS Implementation Specification [18] and geodata servers complying with OGC's Web Feature Service Implementation Specification [19]. However, they focus on building systems that leverage existing ORDBMS and rely on their query engines. Thus, they have little influence on the query engine and they do not offer one that is deliberately customized to the typical query load described in Section 2. Also research projects like the GDI NRW Testbed [3] are more concerned with designing the overall architecture and interaction patterns than with optimizing the underlying data structures.

In [32], the implementation of an OGC CS using a Grid MCS is investigated. While the authors do not address the internal implementation of the service, it emphasizes the importance and impact of our work. Recent research in the MCS area is concerned with managing extensible sets of arbitrary attributes [6], but not with optimizing index structures or exploiting class hierarchies. In [7], the authors describe how to leverage class hierarchies in ebXML Registries, but the authors do not detail on an efficient implementation or index support.

IBM's Cloudscape and Oracle's TimesTen main memory databases are pure relational systems and do not offer any support for type hierarchies or spatial indexes. MonetDB allows for geographic extensions [4], but still has no support for hierarchical relationships. Current research in this area focuses on minimizing CPU branch mispredictions [21] and developing CPU cache conscious data structures [13] to alleviate the main memory access bottleneck. This research is complementary to ours.

6.1 Spatial indexes

A detailed overview on the most important spatial index structures is given in [8]. As we focus on building a query engine rather than on developing a new index structure we pick an existing index structure that assumedly best fits out requirements of performing well with real data sets and coping with high update rates. We chose the MX-CIF quadtree for meeting these requirements and being simple to implement. We do not recommend the best spatial index structure, but we analyze how to utilize any existing index structure to combine spatial dimension and type dimension.

In [29], several approaches to combine spatial indexes and text indexes are presented in order to build a geographical search engine on the web. They aim at enhancing the search precision and recall. However, in the text part of web pages there are considerably more distinct words than we have types in our type hierarchy. No hierarchical relationships between the words are defined or exploited. Finally, their data is stored in files on disk.

6.2 Object-oriented databases

In the area of object-oriented databases (OODB) related indexing problems have been discussed [10, 12, 14, 20]. Indexes combine an object's type with one or several of its attributes. However, attributes may only have a single value instead of a range as in the spatial dimension. Therefore, only point access methods are considered whereas we need to deal with spatially extended geometries requiring spatial access methods. We provide an extensive analysis using real (spatial) data sets. Also, we consider the effort to accommodate changes in the index in order to get a total-cost-of-ownership assessment of the performance in real usage scenarios.

The Multikey type (MT) index [14] is basically similar to our Real 3D index approach using the type hierarchy based mapping variant. However, [14] concentrates on providing an optimal linearization for type hierarchies having multiple inheritance. Scaling the type dimension is not discussed at all, which proves to have a significant impact in our experiments.

6.3 Object-relational databases

Conceptually, in an ORDBMS each type corresponds to a table and each attribute to a column. While some systems (e.g., PostgreSQL) store objects as a row in the table for its type, other systems (e.g., DB2) store all objects in a single hierarchy table [5] which has an additional column for storing the type of an object. In the first case, many tables have to be queried if the sought type is a non-leaf type in the type hierarchy. In the latter case, DB2 internally creates a two-dimensional index on a given column and on the type column. However, it uses a point access method to accomplish this combination, which is inadequate as we have discussed in Section 6.2.

If we have an explicit type column and separate indexes on this column and the geometry column then we can either fetch the qualifying tuple IDs from both indexes and intersect these sets before fetching the remaining object data [1]. Alternatively, we can pick the more selective index to determine a set of candidates, and filter them afterwards [25]. Our experiments have shown, that the latter approach, which we address in Section 4.1, is always more efficient in main memory.

7 Conclusion

In this paper, we have presented a main memory location-conscious query engine that exploits the characteristics of typical queries, which contain spatial predicates and predicates restricting the type of the data. The query engine can be deployed to many components in a large-scale information system and contribute to let the whole system be usable interactively. Both reasons advocate for a main memory approach.

As our experiments have shown, the Real 3D Index approach offers the best overall performance. However, it is crucial to determine an adequate type mapping range or otherwise performance will degrade considerably. We have investigated three different variants to map type IDs to values in the type dimension. The variant relying of object distribution statistics offers the best performance, however the statistics have to be collected beforehand. Both aspects have not been discussed previously. In the usage scenarios with a high type selectivity factor the Real 3D Index approach beats the Separate Indexes approach, which uses conventional database technology, by about 20%. If the type selectivity is low, then the lead increases to almost an order of magnitude. However, the Separate Indexes approach uses only two thirds of the memory and builds up its index at least twice as fast compared to the Real 3D Index approach.

In future work, we will investigate the deployment of the query engine to all components of our data and service provisioning platform for context-aware applications and assess its impact on the overall system. We intend to optimize the overall processing performance under changing workloads (mobile users) and changing data (high level context information derived from sensor data). Thus, our main memory query engine paves the way for virtualizing the whole query processing task by facilitating the distribution of query capabilities across several components based on resources, load, cache contents, etc. Additionally, we plan to extend the index by further dimensions such as valid time or measurement time.

References

1. Astrahan MM, Chamberlin DD: Implementation of a Structured English Query Language. Communications of the ACM, 18(10), 1975
2. Barry DK: Web Services and Service-Oriented Architectures. Morgan Kaufmann Publishers, 2003
3. Bernard L: Experiences from an implementation Testbed to set up a national SDI. 5th AGILE Conf. on Geographic Information Science, Palma, Spain, 2002
4. Boncz PA, Quak W, Kersten ML: Monet And Its Geographic Extensions: A Novel Approach to High Performance GIS Processing. Proc. of the 5th Intl. Conf. on Extending Database Technology (EDBT), Avignon, France, 1996
5. Carey M et al.: O-O, What Have They Done to DB2? 25th Intl. Conf. on Very Large Data Bases (VLDB), 1999
6. Deelman E et al.: Grid-Based Metadata Services. 16th Intl. Conf. on Scientific and Statistical Database Management (SSDBM), Santorini Island, Greece, 2004
7. Dogac A, Kabak Y, Laleci GB: Enhancing ebXML Registries to Make them OWL Aware. Journal on Distributed and Parallel Databases, 18(1), July 2005
8. Gaede V, Günther O: Multidimensional Access Methods. ACM Computing Surveys, 30(2), June 1998
9. Grossmann M et al.: Efficiently Managing Context Information for Large-scale Scenarios. 3rd IEEE Conf. on Pervasive Computing and Communications (PerCom), Kauai Island, Hawaii, March 8-12, 2005
10. Kim W, Kim KC, Dale A: Indexing techniques for object-oriented databases. In: Kim W, Lochovsky FH (eds.): Object-Oriented Concepts, Databases, and Applications. Addison-Wesley, 1989
11. Kolbe TH, Gröger G, Plümer L: CityGML – Interoperable Access to 3D City Models. 1st Intl. Symp. on Geo-Information for Disaster Management (GI4DM), Delft, The Netherlands, 2005
12. Low CC, Ooi BC, Lu H: H-trees: A Dynamic Associative Search Index for OODB. ACM SIGMOD Intl. Conf. on Manangement of Data, San Diego, California, 1992
13. Manegold S, Boncz PA, Kersten ML: Optimizing database architecture for the new bottleneck: memory access. VLDB Journal, 9(3), Dec 2000
14. Mueck TA, Polaschek ML: A configurable type hierarchy index for OODB. VLDB Journal, 6(4), 1997
15. Nebert D: Information Architecture of a Clearinghouse. WWW Conference, 1996, http://www.fgdc.gov/publications/documents/clearinghouse/clearinghouse1.html

16. Nicklas D, Mitschang B: On building location aware applications using an open platform based on the NEXUS Augmented World Model. Software and Systems Modeling, Vol. 3(4), 2004
17. Nicklas D et al.: Design and Implementation Issues for Explorative Location-based Applications: the NexusRallye. VI Brazilian Symposium on GeoInformatics (GeoInfo), 2004
18. Open GIS Consortium: Catalogue Service Implementation Specification. Version 2.0.1, Document 04-021r3, 2004-08-02
19. Open GIS Consortium: Web Feature Service Implementation Specification. Version 1.1, Document 04-094, 2005-05-03
20. Ramaswamy S, Kanellakis PC: OODB Indexing by Class-Division. ACM SIGMOD Intl. Conf. on Manangement of Data, San Jose, 1995
21. Ross KA: Selection Conditions in Main Memory. ACM Trans. on Database Systems, 29(1), March 2004
22. Roubtsov V: My kingdom for a good timer! Reach submillisecond timing precision in Java. JavaWorld, 2003, http://www.javaworld.com/javaworld/javaqa/2003-01/01-qa-0110-timing.html
23. Samet H: The Design and Analysis of Spatial Data Structures. Addison-Wesley, Reading, MA, 1990
24. Schwarz T et al.: Efficient Domain-Specific Information Integration in Nexus. Proc. of the 2004 VLDB Workshop on Information Integration on the Web (IIWeb), Toronto, Canada, 2004
25. Selinger PG et al.: Access Path Selection in a Relational Database Management System. ACM SIGMOD Intl. Conf. on Management of Data, Boston, Massachusetts, 1979
26. Singh G et al.: A Metadata Catalog Service for Data Intensive Applications. ACM/IEEE Conf. on Supercomputing (SC), Phoenix, Arizona, 2003
27. U.S. Census Bureau: TIGER/Line Files. http://www.census.gov/geo/www/tiger/
28. U.S. Census Bureau: TIGER/Line Files Technical Documentation. April 2002, http://www.census.gov/geo/www/tiger/tigerua/ua2ktgr.pdf
29. Vaid S, Jones CB, Joho H, Sanderson M: Spatio-textual Indexing for Geographical Search on the Web. 9th Intl. Symp. on Spatial and Temporal Databases (SSTD), Angra dos Reis, Brazil, 2005
30. Van den Bercken J et al.: XXL - A Library Approach to Supporting Efficient Implementations of Advanced Database Queries. 27th Intl. Conf. on Very Large Data Bases (VLDB), Roma, Italy, 2001
31. Vivid Solutions: JTS Topology Suite. http://www.vividsolutions.com/jts/jtshome.htm
32. Zhao P et al.: Grid Metadata Catalog Service-Based OGC Web Registry Service. 12th annual ACM Intl. workshop on Geographic information systems (ACM GIS), Washington DC, 2004

Approximate Query Processing in Spatial Databases Using Raster Signatures

Leonardo Guerreiro Azevedo[1], Geraldo Zimbrão[1],
Jano Moreira de Souza[1, 2]

[1]Departamento de Ciência da Computação
Universidade Federal do Rio de Janeiro
[2]Departamento de Ciência da Computação, Instituto de Matemática,
Universidade Federal do Rio de Janeiro

1 Introduction

A main issue in database area is to process queries efficiently so that the user does not have to wait a long time to get an answer. However, there are many cases where it is not easy to accomplish this requirement. In addition, a fast answer could be more important for the user than receiving an accurate one. In other words, the precision of the query could be lessened, and an approximate answer could be returned, provided that it is much faster than the exact query processing and it has an acceptable accuracy.

Approximate query processing has emerged as an alternative for query processing in environments for which providing an exact answer can demand a long time. The goal is to provide an estimated response in orders of magnitude less time than the time to compute an exact answer, by avoiding or minimizing the number of disk accesses to the base data [12].

There are a large set of techniques for approximate query processing available in different research areas, as presented by [3]. Good surveys of techniques for approximate query processing are presented by [5, 17]. However, most of the techniques are only suitable for relational databases. On the other hand, providing a short time answer to queries becomes a bigger challenge in spatial database area, where the data usually have high complexity and are available in huge amounts. Furthermore, this subject is a hot research issue in spatial-temporal databases as pointed by [25].

Spatial data consists of spatial objects made up of points, lines, regions, rectangles, surfaces, volumes, and even data of higher dimension which includes time [26]. Examples of spatial data include cities, rivers, roads, counties, states, crop coverage, mountain ranges etc. It is often desirable to attach spatial with non-spatial attribute information. Examples of non-spatial data are road names, addresses, telephone numbers, city names etc. Since spatial and non-spatial data are so intimately connected, it is not surprising that many of the issues that need to be addressed are in fact database issues.

Spatial DBMS (Database Management Systems) provides the underlying database technology for Geographic Information Systems (GIS) and other applications [13]. There are numerous applications in spatial database systems area, such as: traffic supervision, flight control, weather forecast, urban planning, route optimization, cartography, agriculture, natural resources administration, coastal monitoring, fire and epidemics control [1, 27]. Each type of application deals with different features, scales and spatiotemporal properties.

In a traditional SDBMS query processing environment (Fig.1.a), user queries are sent to the database that processes them and returns to the user an exact answer. On the other hand, in a SDBMS set-up for providing approximate query answers, a new component is added, the approximate processing engine (Fig.1.b). In this new framework, queries are sent directly to the approximate processing engine. It processes the query and returns an approximate answer to the user, along with a confidence interval showing the response accuracy. If the precision is not sufficient for the user to take his decision, the query can be processed by the SDBMS, providing an exact answer to the user.

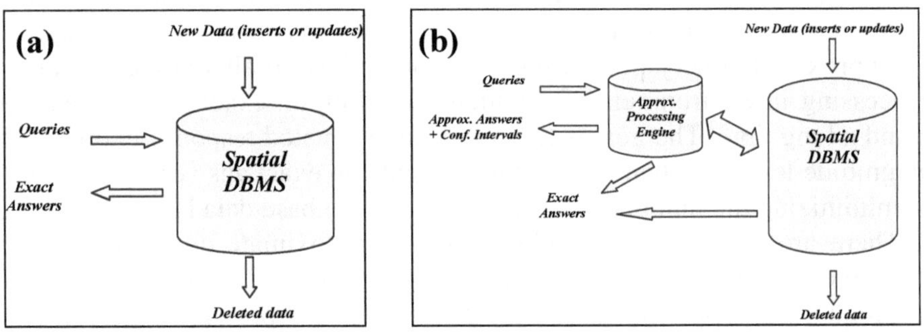

Fig. 1. SDBMS query processing environment: (a) traditional environment (b) set-up for approximate processing

The approximate processing engine stores reduced representation of real data to perform the approximate processing. Therefore, it is also possible, to execute the query partially over approximate data and partially over real data, when it is not guaranteed that some approximations (or synopses) of objects do not produces the desired accuracy or if the real representation of objects are so simple that the processing costs of computing the exact or the approximate answer are the same. For instance, to compute the area of a polygon of few vertices could be executed in almost the same time as the time to process the same query over a synopsis of the real data.

One important issue regards to the approximate processing engine is related to the maintenances of data. When new data arrives or stored data is updated, it is also required to store the representations of new objects in the engine or update the existing synopses. Therefore, it is also important that synopses can be computed quickly. In the case of deletion of objects, the existing synopses must be deleted as well. Thus all information that comes or leaves the SDBMS must be sent to the approximate engine in order to keep it up-to-date.

This work is concerned with the use of raster signatures in approximate query processing in spatial databases. We extended the proposals of [3, 4] of using Four-Color Raster Signature (4CRS) [28] for fast and approximate processing of queries over polygon datasets. We propose several new algorithms for a set of spatial operations that can be processed approximately using 4CRS. 4CRS stores the main data characteristics in an approximate and compact representation that can be accessed and processed faster than real data can be accessed and processed. By doing so, the exact geometries of objects are not processed during query execution, which is the most costly step of a spatial operation, since it requires to search and to transfer large objects from disk to main storage [7, 21]. Also, the exact processing needs to use CPU-time intensive algorithms to decide whether the objects match the query condition [6]. As a result, the approximate query answer is returned in a shorter time than the time to return an exact answer. On the other hand, the answer is estimated and not exact, so a precision measure is also returned as a confidence interval that allows the user to decide if the accuracy of the response is sufficient. In general, this approximate answer is enough for many applications.

It is important to emphasize that the main target of this work is to present our proposals of algorithms, while the experimental evaluation of them is a current work being developed on Secondo [14], which is an extensible DBMS platform for research prototyping and teaching. However, the experimental results of evaluating approximate processing against exact processing presented by [3, 4] demonstrated the efficiency of using 4CRS for approximate query processing. [3] evaluated an algorithm for computing polygon approximate area and an algorithm for computing the approximate area of polygon × window intersection (window query). According to those evaluations, the approximate processing is 3.5 times and 14 times faster than the exact processing in response time, while 52 times and 14 times faster relate to number of disk accesses, respectively. The response error is also quite small, an average of -2.62% and 1%, respectively. [4] evaluated an algorithm for estimating the overlapping area of polygon join. In this case, the approximate processing varies from 5 to 15 times faster than the exact processing in response time and from 5 to 10 times faster related to number of disk accesses. Approximate answers have an average error of 0.6%.

This work is divided in sections, as follows: Section 2 presents scenarios and applications where approximate query processing can be used; Section 3 presents our proposals of algorithms; and, in Section 4, is dedicated to conclusions and future work.

2 Scenarios and Applications

There are many scenarios and applications where a slow exact answer can be replaced by a fast approximate one, provided that it has the desired accuracy. [18] emphasizes that in Decision Support Systems the increasing in business competitiveness is requiring an information-based industry to make more use of its accumulated data, and thus techniques of presenting useful data to decision makers in a timely manner are becoming crucial. They propose also the use of approximate query processing during a drill-down query sequence in ad-hoc data mining, where the earlier queries in the sequence are used solely to determine what the interesting queries are. [24] proposes to use approximate query processing for spatial OLAP. [12] highlights that an approximate answer can provide feedback on how well-posed a query is.

[19, 12] propose the use of approximate query processing to define most efficient execution plan for a given query while [8] propose its use in selectivity estimation in Spatial Database Management Systems (SDBMS) in order to return approximate results that come with provable probabilistic quality guarantees.

An approximate answer can also be used as an alternative answer when the data is unavailable in data warehousing environments and in distributed data recording as pointed by [10, 12, 20] or in mobile computing as highlighted by [22].

[9] indicates to use approximate query processing in order to make decisions and infer interesting patterns on-line, such as over continuous data streams.

3 Approximate query processing using Four-Color Raster Signature

The 4CRS signature was first used to improve the processing of spatial joins of polygon datasets. It was proposed by [28] as a filter in the second step of the Multi-Step Query Processor (MSQP) [7] in order to reduce exact geometry test of spatial objects. 4CRS is a polygon raster signature represented by a small bit-map of four colors upon a grid of cells. Each grid cell has a color representing the percentage of the polygon's area within cell: *Empty* (0% of intersection); *Weak* (the cell contains an intersection of 50% or less with the polygon); *Strong* (the cell contains an intersection of more than 50% with the polygon and less than 100%); and, *Full* (the cell is fully occupied by the polygon). The grid can have its scale changed in order to obtain a more accurate representation (higher resolution) or a more compact one (lower resolution). Further details of 4CRS signature can be found in [3, 28].

The 4CRS characteristics and the good results obtained using 4CRS as geometric filter in polygon join processing motivated its use on approximate query processing. This new approach has the goal not solely to reduce the number of objects that have their exact geometry processed. Instead, we propose to return to the user an approximate answer that is obtained processing the query over the 4CRS signatures of polygons, without accessing the object's real representation, and not executing the exact geometry step (the most expensive one). Hence, new algorithms must be designed, implemented and evaluated to concern to the requirements of this new of approach.

3.1 Approximate Operations

There are many operations that could benefit from a fast and approximate query processing, so that the user could have an answer in a short time instead of waiting a long time for an exact answer. In this work we present our proposals of approximate query processing algorithms based on the classification proposed by [15, 16] in the Rose Algebra. They divide the spatial operations into four groups:

- **Spatial operators returning numbers**: area, number of components, distance, diameter, length, perimeter;
- **Spatial predicates**: equal, different, disjoint, inside, area disjoint, edge disjoint, edge inside, vertex inside, intersects, meet, adjacent, border in common;
- **Operators returning spatial data type values**: intersection, plus, minus, common border, vertices, contour, interior;
- **Spatial operators on set of objects**: sum, closest, decompose, overlay, fusion.

In the next sections, because of the limited space, we present our proposals of algorithms for the most important operations for understand our new approach. In order to make the descriptions simpler, we present similar operations in the same section.

3.1.1 Approximate Area of Polygon

The algorithms that return the approximate area of polygon and the approximate area of polygon within cell are proposed by [3], while the algorithm that returns the approximate area of polygon join are proposed by [4], we extended those proposals in this work. The first two algorithms are based on the expected area of polygon within cell and the last is based on the expected area of intersection of two types of cells. These definitions are presented in details in those works. Therefore, we will present here only short explanations of those definitions, which are used to describe our new proposals of algorithms.

The expected area of polygon within cell corresponds to a sum of estimatives of the area of polygon within cell types. For example, an expected area of polygon within an *Empty* cell is equal to 0% (zero percent), since there is no portion of the polygon inside the cell. For *Weak*, *Strong* and *Full* cells the estimatives are 25%, 75% and 100%, respectively.

The expected area of intersection of two types of cells is used to estimate the intersection of two polygons, which is approximately answered as the intersection of their 4CRS signatures. For example, the expected area corresponding to a combination of an *Empty* cell with any other type of cell results in an expected area of 0% (zero percent). Similarly, when two *Full* cells overlap, the expected area is 100%. More details about expected area are presented in [3, 4].

3.2.2 Distance and Diameter

The distance between two polygons can be estimated from their 4CRS signatures, computing the distance among cells corresponding to polygons' borders (*Weak* and *Strong* cells). The result can be estimated as the average of the minimum and maximum distances computed as follow. The minimum distance is the distance between the outer borders of cells (borders adjacent to borders of *Empty* cells), while the maximum distance is computed from the inner borders of these cells (i.e., borders opposite to the outer borders). The minimum and maximum distances can be used to define a confidence interval for the computed approximate distance. Fig.2 presents an example of computing the distance between two polygons using their 4CRS. The polygons are presented in Fig.2.a, while Fig.2.b shows their 4CRS signatures. Fig.2.c is a zoom of a cell combination used to compute the approximate distance, highlighting the minimum and maximum distances between two cells.

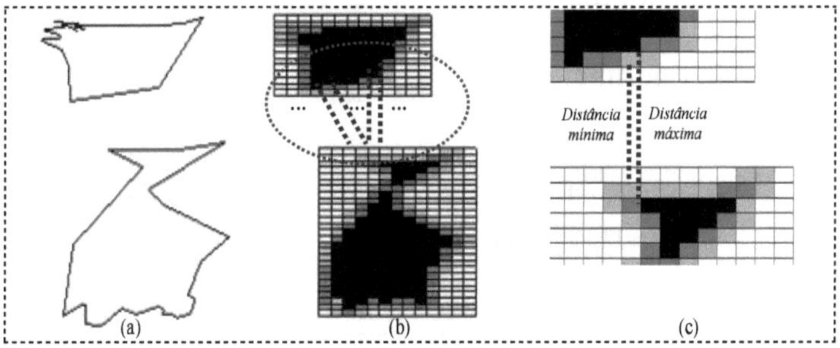

Fig. 2. Example of computing the minimum and maximum distance between two polygons from their 4CRS signatures

The diameter of a spatial object is defined as the longest distance between any of its components. Therefore, in the case of polygons, the diameter is the longest distance between the faces that compose the polygon. So, the diameter can be computed using the same algorithm to compute distance between polygons, since each face has a different 4CRS signature.

3.1.3 Perimeter and Contour

The perimeter operation calculates the sum of the length of all cycles of a region (or polygon) value. If we intend to compute only the sum of the length of the outer cycles and not consider the holes of a polygon, the contour operator can be used to eliminate holes first. The perimeter of a polygon can be computed from its 4CRS signature. A simple proposal is to compute the perimeter as the average of the outer perimeter and the inner perimeter. Fig.3 presents an example of outer and inner perimeters of a polygon. Fig.3.a and Fig.3.b show the polygon and its 4CRS signature, respectively. Fig.3.c presents the outer perimeter of the 4CRS signature, and Fig.3.d presents the inner perimeter. The outer perimeter can be computed as the sum of the length of edges corresponding to the borders of cells of type different from *Empty*, those are adjacent to *Empty* cells (for example, cell c' presented in Fig.3.b, whose *Top* and *Left* edges would be considered as part of the outer perimeter) or adjacent to the border of the signature's MBR (for example, cell c'' presented in Fig.3.b, whose *Top* and *Left* edges would also be considered as part of the outer perimeter). MBR (Minimum Bounding Rectangle) is the smallest rectangle that encloses a spatial object. It is the most popular geometric key. MBR reduces a spatial object's complexity to four parameters that retain its most important characteristics: position and extension. The signature's MBR is the smallest rectangle that encloses the object's signature, and its coordinates are in power of two [28]. On the other hand, the inner perimeter could be computed as the sum of length of edges of cells adjacent to the cells that were considered when computing the outer perimeter.

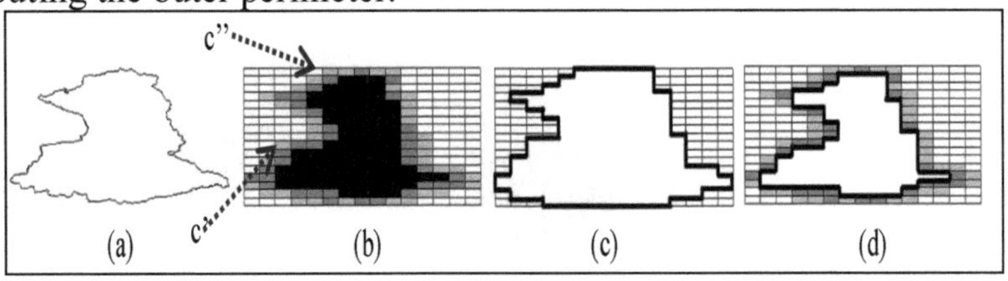

Fig. 3. Example of outer and inner perimeter of polygon from its 4CRS signature

The calculus of the contour of a polygon can be approximately computed from its 4CRS signature similarly to the calculus of the perimeter of a polygon. We can connect the medium points of *Weak* and *Strong* cells in order to compute the contour of the polygon, and we can assume that the maximum contour corresponds to the cells that compose the outer perimeter (Fig.3.c) and the minimum contour corresponds to cells that compose the inner perimeter (Fig.3.d).

3.1.4 Equal and Different

In exact processing, the equal and different operations return exactly if two objects are equal or not, respectively. On the other hand, in approximate processing using 4CRS is not always possible to state that objects are equal or different with 100% of confidence. In this case, the equality and inequality is defined as a function that returns a value in the interval [0,1] that indicates a true percentage of the equality (or inequality) of objects. We call this value as "affinity degree". By doing so, we can test if two objects are equal (or different) using their 4CRS signatures. For each comparison of pair of cells we compute an affinity degree. The final affinity degree is equal to the sum of the individual degrees divided by the number of comparisons of pair of objects, if no trivial case occurs. So, for instance, when comparing a pair of *Empty* cells, we can state that the polygons are 100% equal, since *Empty* cells do not have any intersection with polygon. Similarly, two polygons are equal when comparing *Full* cells. On the other hand, when comparing *Weak* and *Strong* cells a different reasoning must be used. Our proposal is to use for these cases the concept of expected area employed by the algorithm that computes the approximate area of polygon × polygon intersection [4]. In other words, for exact cases (comparisons of *Empty* ×*Empty* and *Full* × *Full* cells) the equality (or affinity degree) is 1, while for other cases the affinity degree is equal to the expected area. For instance, the overlap of *Weak* × *Weak* cells contributes with 0.0625, while the overlap of *Strong* × *Strong* cells contributes with 0.5625, which represent the intersection of polygon intersection using their 4CRS signatures' cells. The answer is computed as the sum of the affinity degrees divided by the number of comparisons. It is important to highlight that if exists at least one overlap of different cell types we can state that the objects are not equal.

```
real equal(4CRS1, 4CRS2)
   if (4CRS1.lengthOfCellSide ≠ 4CRS2.lengthOfCellSide)or
      (4CRS1.nCells ≠ 4CRS2.nCells)or (4CRS1.mbr ≠ 4CRS2.mbr) then
      return 0;
   affinityDegree = 0; nRuns=0;
   for each c1 cell of 4CRS1 do
     for each c2 cell of 4CRS2 that overlaps c1 do
       if c1.type==c2.type then
          if c1.type==Empty or c1.type==Full then
             affinityDegree += 1;
          else if c1.type==Weak then
             affinityDegree += 0.0625;
          else
             affinityDegree += 0.5625;
       else
         return 0;
       nRuns++;
  return affinityDegree / nRuns;
```

Code. 1. Algorithm that returns if two polygons are approximately equal

The algorithm that returns if two objects are equal is presented in Code 1. The algorithm returns 0 if the polygons are not equal, otherwise it returns an affinity degree that shows a measure of equality of the objects.

In the case of different operation, we can use a similar algorithm, replacing the return values and affinity degrees by their complements. For instances, we replace "return 0" by "return 1" and we replace "affinityDegree += 0.0625" by "affinityDegree += 1 - 0.0625".

3.1.5 Disjoint, Area Disjoint, Edge Disjoint

Two objects are disjoint if they have no portion in common. In the case of area disjoint, the objects do not have area in common, but they can have edges overlap. On the other hand, two objects are edge disjoint if they do not have overlap of edges. 4CRS signatures can be used to estimate if two objects are disjoint, area disjoint or edge disjoint. In some cases it is also possible to return an exact answer.

When comparing the 4CRS signatures of two polygons, if there are only overlap of *Empty* cells × any other cell type, we can state that the polygons are disjoint, and, consequently, they are also area disjoint and edge disjoint. On the other hand, if there is at least one overlap of a *Full* cell × *Weak* or *Strong* or *Full* cell, we can state that the objects are not disjoint nor area disjoint nor edge disjoint. It is also possible to state that two polygons are edge disjoint if there is no overlap of *Weak* × *Strong* cells, i.e., or one polygon is inside the other, or it is outside the other. Therefore, an approximate answer is returned to the user only when there are intersections of *Weak* × *Strong* cells, otherwise we can return an exact answer. Our proposal is to use the expected area of intersection of two types of cells to estimate the answer for those cases, and to return to the user an affinity degree of the answer. Thus, a weight of 100% is assigned to comparisons of pair of cells where it is possible to have an exact answer. In the cases that an approximate value is computed, we propose to use the complement value of the expected area of intersection of two types of cells, since we are interested in estimating disjunction of polygons, which is opposite to estimating the intersection area of polygons.

Code 2 presents an algorithm to compute an affinity degree about disjunction of two polygons. This algorithm can be also used to determine if two objects are area disjoint.

The algorithm to evaluate if two polygons are edge disjoint is very similar to the algorithm proposed in Code 2. The only difference is in the part that compares *Full* × *Full* cells that must be replaced by "affinityDegree += 1".

```
real disjoint(4CRS1, 4CRS2)
  interMBR = intersectionMBR(4CRS1, 4CRS2);
  if interMBR is NULL then /*Does not exist MBR intersection*/
     return 1;
  /*change scale if it is needed*/
  changeScale(4CRS1, 4CRS2);
  affinityDegree = 0; nRuns = 0;
  for each c1 cell from 4CRS1 that is inside interMBR do
     for each c2 cell from 4CRS2 that intersects c1 do
        if ( (c1.type == Empty) or (c2.type == Empty) ) then
           affinityDegree += 1;
        else
           if ( (c1.type == Weak) and
              (c2.type == Weak or c2.type == Strong) ) or
              ( (c1.type == Strong) and (c2.type == Weak)) then
              affinityDegree += (1 - expectedArea[c1.type,c2.type]);
           else /*Full × Full*/
              return 0;
     nRuns++;
  return affinityDegree / nRuns;
```

Code.2. Algorithm for returning if two objects are disjoint

3.1.6 Inside (Encloses), Edge Inside, Vertex Inside

From the 4CRS signatures of two polygons is possible to state that a polygon P_1 is inside a polygon P_2 if all cells of 4CRS signature of P_1, different from *Empty*, are overlapped by *Full* cells of the 4CRS signature of P_2. On the other hand, if there is an overlap of at least P_1 cell of type different from *Empty* with an *Empty* cell of P_2, we can state that P_1 is not inside P_2. In the case of overlap of *Weak* × *Strong* cells or *Weak* × *Weak* cells or *Strong* × *Strong* cells, it is not possible to return an exact answer. Hence we need to define an approximate value for these cases of cell overlaps. We propose to use the expected area of intersection of two types of cells for estimating the answer. The algorithm for inside and edge inside operations are the same. However, the vertex inside operation cannot be approximately processed using 4CRS signature, since information about polygon's vertices is not stored by the signature. It is possible to return if a polygon is not vertex inside related to other polygon when they do not intersect, or that the polygon is vertex inside when it is completely inside the other polygon. On the other hand, when the polygons intersect, but one polygon is not inside the other, it is not possible to return an approximate answer for this operation using 4CRS signatures.

Code 3 proposes an algorithm for returning if a polygon P_1 is inside other polygon P_2, comparing their 4CRS signature. It is important to note that, according to the algorithm that computes the grid of cells [28], if the cell size of signat4CRS1 is greater than the cell size of signat4CRS2, polygon P_1 represented by assinat4CRS1 is bigger than polygon P_2 represented by assinat4CRS2. Hence P_1 is not inside P_2.

```
real inside(4CRS1, 4CRS2)
 if (4CRS1.lengthOfCellSide > 4CRS2.lengthOfCellSide) then
     return 0;
  /*change scale if it is needed*/
  changeScale(4CRS1, 4CRS2);
  interMBR = intersectionMBR(4CRS1, 4CRS2);
  if interMBR is NULL then /*Does not exist MBR intersection*/
     return 0;
  affinityDegree = 0;
  for each c1 cell from 4CRS1 that is inside interMBR do
    for each c2 cell from 4CRS2 that intersects c1 do
       if (c1.type == Empty and c2.type • Empty) then
          return 0;
       else
         if (c1.type == Weak or c1.type == Strong) then
           if (c2.type == Weak or c2.type == Strong) then
             affinityDegree += expectedArea[c1.type, c2.type];
           else
             if (c2.type == Empty) then
                 affinityDegree += 1;
             else /*s.type == Full*/
                return 0;
         else /*c1.type == Full x any c2.type*/
           affinityDegree += 1;
       nRuns++;
  return affinityDegree / nRuns;
```

Code.3. Algorithm for returning if a polygon P_1 is inside other polygon P_2, according to their 4CRS signature

3.1.7 Intersects and Intersection

While the intersect operation returns if two polygons intersect, the intersection operation returns the polygon resulting from the intersection.

For the intersect operation an affinity degree is returned. This value is computed using the expected area of intersection of two types of cells. Code 4 presents an algorithm to evaluate if two polygons intersect. It is important to highlight that in many cases it is possible to return an exact answer.

In the case of the algorithm that returns the polygon resulting from the intersection of two polygons evaluating their 4CRS signatures, we propose the following approach: to create a new 4CRS signature from the intersection of the 4CRS signatures of the polygons, and generate a polygon connecting the medium points of the border cells of the new signature (*Weak* and *Strong* cells). The cell types of the new signature can be set according to the following values: if the value resulting from the intersection of the pair of cells is in the interval (50%, 100%) then the type of the new cell is *Strong*; and, if the value is in the interval (0%, 50%] the type of the new cell is *Weak*. 0% and 100% of intersection define cell types equal to *Empty* and *Full*, respectively. It is important to highlight that, along with the polygon that represents the intersection, an affinity degree is also returned in order to show a degree of similarity between the approximate polygon and the polygon that would represent the exact intersection.

```
real intersects(4CRS1, 4CRS2)
  interMBR = intersectionMBR(4CRS1, 4CRS2);
  /*change scale if it is needed*/
  changeScale(4CRS1, 4CRS2);
  affinityDegree = 0;
  for each c1 cell from 4CRS1 and c2 cell from 4CRS2
      that intersects and are inside interMBR do
    if (c1.type == Full) and (c2.type • Empty) then
        return 1;
    else
        if ( (c1.type == Strong) and
           ( (c2.type==Strong) or (c2.type==Full) ) ) then
           return 1;
        else
           if ( (c1.type == Weak) and (c2.type==Full)) then
              return 1;
           else
              if ( (c1.type == Empty) and (c2.type==Empty) ) then
                 affinityDegree += 1;
              else
                affinityDegree += expectedArea[c1.type,c2.type];
    nRuns++;
  return affinityDegree / nRuns;
```

Code. 4. Algorithm to evaluate if two polygons intersect

3.1.8 Adjacent, Border in Common, Common border

Two polygons are adjacent if they have at least a portion of their borders in common, which is quite similar to evaluate if two polygons have a border in common. Thus, we are proposing to use the same algorithm that returns an approximate answer if two polygons are border in common and to return if they are adjacent.

One proposal of operation that returns if two polygons are border in common is to employ the expected area of intersection of two types of cells, in order to return an affinity degree as the answer. Polygon borders are composed by segments; hence, they do not have area. Therefore, a common border of two polygons can be found only on the overlap of *Weak* or *Strong* cells, which are the cells where the borders are. Thus, it is possible to return an exact answer when there is no overlap of these types of cells. On the other hand, an approximate value is returned. The algorithm is presented in Code 5.

```
real borderInCommon(4CRS1, 4CRS2)
  interMBR = intersectionMBR(4CRS1, 4CRS2);
  if interMBR is NULL then /*Does not exist MBR intersection*/
    return 0;
  /*change scale if it is needed*/
  changeScale(4CRS1, 4CRS2);
  affinityDegree = 0; nOverlaps = 0;
  for each c1 cell from 4CRS1 that is inside interMBR do
     for each c2 cell from4CRS2 that is inside c1 do
        if (c1.type == Weak or c1.type == Strong) then
           nOverlaps += 1;
           if (c2.type == Weak or c2.type == Strong) then
              affinityDegree += expectedArea[c1.type,c2.type];
        else
           if (c2.type == Weak or c2.type == Strong) then
              nOverlaps += 1;
              if (c1.type == Weak or c1.type == Strong) then
                 affinityDegree += expectedArea[c1.type,c2.type];
  return affinityDegree / nOverlaps;
```

Code.5. Algorithm to return if two polygons have a border in common

One proposal of algorithm to return an approximate answer as the border in common of two polygons is to adapt the algorithm presented in Code 5 in order to create segments connecting the medium points of the overlap of *Weak* × *Weak* cells, *Weak* × *Strong* cells and *Strong* × *Strong* cells. An affinity degree can be returned using the same idea as presented in the algorithm proposed in Code 5.

3.1.9 Plus and Sum

The plus operator computes the union of two objects, while the sum operator computes the union of a set of objects.

If two polygons do not have MBR intersection, then the polygon that represents the union of these objects is composed by the faces of these two polygons. On the other hand, when the polygons have MBR intersection, a new 4CRS signature is created, according to the algorithm presented in Code 6. A new polygon is computed from the 4CRS generated signature, connecting the medium points of *Weak* and *Strong* cells (function *computePolygon*).

```
Polygon union(4CRS1, 4CRS2)
   interMBR = intersectionMBR(4CRS1, 4CRS2);
   if interMBR is NULL then /*Does not exist MBR intersection*/
       polygon1 = computePolygon(4CRS1);
       polygon2 = computePolygon(4CRS2);
       polygon.addFaces(polygon1);
       polygon.addFaces(polygon2);
       return polygon;
   /*change scale if it is needed*/
   changeScale(4CRS1, 4CRS2);
   unionMBR = computeUnionMBR(4CRS1.mbr,4CRS2.mbr);
   /*Create 4CRS signature with only Empty cells*/
   n4CRS = createSignature(unionMBR, 4CRS1.lengthOfCellSide, Empty);
   for each c1 cell from 4CRS1 that intersects n4CRS cell n do
      n.type = c1.type;
      for each c2 cell from 4CRS2 that intersects n4CRS cell n do
         if n.type == Empty or c2.type == Full then
            n.type = c2.type;
         else if n.type == Weak and c2.type == Strong then
            n.type = c2.type;
   polygon = computePolygon(n4CRS);
   return polygon;
```

Code. 6. Algorithm to compute the union of two polygons using their 4CRS signatures

3.1.10 Minus

The minus operator applied on polygon P_1 related to polygon P_2 is composed by the portion of P_1 that does not have intersection with P_2. One proposal for this operation using their 4CRS is to set to *Empty* the cells of P_1 signature that is overlapped by cells of types *Strong* and *Full* of P_2 signature. In order to compute the polygon from the resulting 4CRS signature we must consider that *Full* cells can be part of the new polygon's border, besides *Weak* and *Strong* cells. The algorithm is presented in Code 7.

```
Polygon minus(4CRS1, 4CRS2)
    interMBR = intersectionMBR(4CRS1, 4CRS2);
    if interMBR is NULL then /*Does not exist MBR intersection*/
      return createPolygon(4CRS1);
    /*change scale if it is needed*/
    changeScale(4CRS1, 4CRS2);
    for each c1 cell from 4CRS1 that is inside interMBR do
       for each c2 cell from 4CRS2 that intersects c1 do
          if c2.type == Strong or c2.type == Full then
             c2.type = Empty;
    polygon = createPolygon(4CRS1, ConsiderAlsoFullCellsAsBorder);
    return polygon;
```

Code.7. Algorithm to compute the minus operation of two polygons using their 4CRS signatures

4 Conclusions

This work proposed new algorithms for approximate query processing in spatial databases using raster signatures. The target was to provide an estimated result in orders of magnitude less time than the time to compute an exact answer, along with a confidence interval for the response. We extended the proposals of [3] and [4] of using Four-Color Raster Signature (4CRS) [28] for fast and approximate processing of queries on polygon datasets. By doing so, the exact geometries of objects are not processed during the query execution, which is the most costly step of the spatial query processing since it requires to search and to transfer large objects from disk to the main storage [7, 21]. Also, the exact processing algorithm needs to use complex intensive CPU-time algorithms for deciding whether the objects match the query condition [6]. There are many scenarios and applications where a slow exact answer can be replaced by a fast approximate one, provided that it has the desired accuracy, as presented in Section 2.

In Section 3 we presented proposals of algorithms for approximate operations using 4CRS, which are the main contributions of this work. The experimental evaluation is not addressed in this work; it is on going work developed on Secondo [14], which is an extensible DBMS platform for research prototyping and teaching.

As future work, we plain to implement and to evaluate algorithms involving other kinds of datasets, for example, points and polylines, and combinations of them, point × polyline, polyline × polygon and polygon × polyline. [2] proposed a raster signature for polylines named Five-Colors Directional Raster Signature (5CDRS). In that work, 5CDRS signature was employed as a geometric filter for processing spatial joins in multiple steps of datasets made by polylines. The results obtained were quite good. Our proposal is to evaluate 5CDRS signature for approximate query processing of polyline datasets. Besides, we also expect to evaluate the 5CDRS and 4CRS signatures together, testing the approximate query processing of queries involving polylines and polygons. The results obtained employing both signatures as geometric filter in multi-step spatial join were also quite good, as presented by [23].

References

[1] Aronoff, S. (1989) Geographic Information Systems, WDL Publications, 1st edn.

[2] Azevedo, L. G., Monteiro, R. S., Zimbrao, G., Souza, J. M. (2003) Polyline Spatial Join Evaluation Using Raster Approximation. In: GeoInformatica, Kluwer Academic Publishers , vol. 7, n. 4, pp. 315-336.

[3] Azevedo, L. G., Monteiro, R. S., Zimbrao, G., Souza, J. M. (2004) Approximate Spatial Query Processing Using Raster Signatures. In: Proceedings of VI Brazilian Symposium on GeoInformatics, Campos do Jordao, Brazil, pp. 403-421.

[4] Azevedo, L. G., Zimbrao, G., Souza, J. M., Güting, R. H. (2005) Estimating the overlapping area of polygon join. In: Proceedings of the 7th International Symposium on Advances in Spatial and Temporal Databases, Angra dos Reis, Brazil, pp. 187-194.

[5] Barbara, D., Dumouchel, W., C. Faloutsos, et al. (1997) The New Jersey data reduction, Bulletin of the Technical Committee on Data Engineering, vol. 20, n. 4, pp. 3-45.

[6] Brinkhoff, T., Kriegel, H. P., Schneider, R (1993) Comparison of Approximations of Complex Objects Used for Approximation-based Query Processing in Spatial Database Systems. In: Proceedings of the Ninth International Conference on Data Engineering, Vienna, Austria, pp. 40-49.

[7] Brinkhoff, T., Kriegel, H. P., Schneider, R., Seeger, B. (1994) Multi-step Processing of Spatial Joins, ACM SIGMOD Record, vol. 23, n.2, pp. 197-208.

[8] Das, A., Gehrke, J., Riedwald, M. (2004) Approximation Techniques for Spatial Data. In: Proceedings of the 2004 ACM-SIGMOD International Conference on Management of Data, Paris, France, pp. 695-706.

[9] Dobra, A., Garofalakis, M., Gehrke, J. E., Rastogi, R. (2002) Processing complex aggregate queries over data streams. In: Proceedings of the 2002 ACM-SIGMOD International Conference on Management of Data, Madison, Wisconsin, USA, pp. 61-72.

[10] Faloutsos, C., Jagadish, H. V., Sidiropoulos, N. D. (1997) Recovering information from summary data. In: Proceedings of 23rd International Confeference on Very Large Data Bases, Athens, Greece, pp. 36-45.

[12] Gibbons, P. B., Matias, Y., Poosala, V. (1997) Aqua project white paper. Technical Report, Bell Laboratories, Murray Hill, New Jersey, USA.

[13] Güting, R. H. (1994), An Introduction to Spatial Database Systems, The International Journal on Very Large Data Bases, vol. 3, n. 4, pp. 357-399.

[14] Güting, R.H., Almeida, V., Ansorge, D., Behr, T., Ding, Z., Höse, T., Hoffmann, F., Spiekermann, M., and Telle, U. (2005) Secondo: An Extensible DBMS Platform for Research Prototyping and Teaching, In Proceedings of 21st International Conference. on Data Engineering (ICDE'05) (Tokyo, Japan, 2005), IEEE Computer Society, Washington, DC, USA, pp. 1115-1116.

[15] Güting, R.H., Schneider, M. (1995) Realm-Based Spatial Data Types: The ROSE Algebra. The International Journal on Very Large Data Bases, vol 4, n. 2, pp. 243 - 286.
[16] Güting, R. H., De Ridder, T., Schneider, M. (1995) Implementation of the ROSE Algebra: Efficient Algorithms for Realm-Based Spatial Data Types. In: Proceedings of the 4th International. Symposium on Large Spatial Databases Systems, Portland, USA, pp. 216-239.
[17] Han, J., Kamber, M. (2001) Data Mining: concepts and techniques, Morgan Kaufmann publishers, 1st edn.
[18] Hellerstein, J. M., Haas, P. J., Wang, H. J. (1997) Online aggregation. In: Proceedings of ACM SIGMOD International Conference on Management of Data, Tucson, Arizona, USA, pp. 171-182.
[19] Ioannidis, Y. E., Poosala, V. (1995) Balancing histogram optimality and practicality for query result size estimation. In: Proceedings of the 1995 ACM SIGMOD international conference on Management of data, San Jose, California, USA, pp. 233-244.
[20] Jagadish, H. V., Mumick, I. S., Silberschatz, A. (1995) View maintenance issues in the chronicle data model. In: Proceedings of the fourteenth ACM SIGACT-SIGMOD-SIGART Symposium on Principles of Database Systems, San Jose, California, USA, pp. 113-124.
[21] Lo, M. L., Ravishankar, C. V. (1996) Spatial Hash-Joins. In: Proceedings of the 1996 ACM-SIGMOD International Conference on Management of Data, Montreal, Quebec, Canada, pp. 247-258.
[22] Madria, S. K., Mohania, M. K., Roddick, J. F. (1998) A Query Processing Model for Mobile Computing using Concept Hierarchies and Summary Databases. In: The 5th International Conference of Foundations of Data Organization, Kobe, Japan, pp. 147-157.
[23] Monteiro, R. S., Azevedo, L.G., Zimbrao, G., Souza, J. M. (2004) Polygon and Polyline Join Using Raster Filters. In: Proceedings of the 9th International Conference on Database Systems for Advances Applications, Jeju Island, Korea, pp. 255-261.
[24] Papadias, D., Kalnis, P., Zhang, J. et al. (2001) Efficient OLAP Operations in Spatial Data Warehouses. In: Proceedings of the 7th International Symposium on Advances in Spatial and Temporal Databases, Redondo Beach, CA, USA, pp. 443-459.
[25] Roddick, J., Egenhofer, M., Hoel, E., *et al.* (2004) Spatial Temporal and Spatiotemporal Databases Hot Issues and Directions for PhD Research. SIGMOD Record, vol. 33, n. 2, pp. 126-131.
[26] Samet, H. (1990) The Design and Analysis of Spatial Data Structure, Addison-Wesley Publishing Company, 1st edition.
[27] Tao, Y., Sun, J., Papadias, D. (2003) Selectivity estimation for predictive spatio-temporal queries. In: Proceedings of the 19th International Conference on Data Engineering, Bangalore, India, pp. 417-428.
[28] Zimbrao, G., Souza, J. M. (1998) A Raster Approximation for Processing of Spatial Joins. In: Proceedings of the 24th International Conference on Very Large Data Bases, New York City, New York, USA, pp. 558-569.

A Rule-Based Optimizer for Spatial Join Algorithms

Miguel Fornari[1,3], João Luiz Dihl Comba[1], Cirano Iochpe[1,2]

[1]Instituto de Informática
Universidade Federal do Rio Grande do Sul
[2]Empresa de Tecnologia da Informação e Comunicação da Prefeitura de Porto Alegre
[3]Faculdade de Tecnologia
SENAC/RS

1 Introduction

The spatial join operation combines two sets of spatial features, *A* and *B*, based on a spatial predicate [21]. Combining such pairs of spatial features in large data sets implies the execution of both Input/Output (I/O) and a large number of CPU operations. Therefore, it is both one of the most important and the most expensive operations in geographic databases systems (GDBMS).

As an example, consider a geographical data set describing rivers and another one representing counties. Many applications, like transport planning, agricultural production and flood prevention, must know which counties are crossed by rivers. To answer the query "find every county that is spatially crossed by a river" a user can apply a spatial join over the two feature sets with the topological predicate "crosses".

Traditionally, a user submits a query to the Database Management System (DBMS), using a high level language, such as SQL. After lexical and syntactic validation, the query is transformed into a relational algebra expression, to be processed by the query optimizer module. The query optimizer, based on a set of statistical data stored in the data dictionary, defines an execution plan. The evaluation engine performs the query according to the execution plan over the user data.

The main contribution of this work consist of a set of rules to optimize the performance of some well-known algorithms: *Partition Based Spatial Merge Join (PBSM)* [19], *Iterative Stripped Spatial Join (ISSJ)* [13], Synchronized *Tree Transversal (STT)* [5] and *Histogram-based Hash Stripped Join (HHSJ)* [8].

The goal is to reduce the response time of spatial join algorithms, changing some basic parameters. We must address two main problems: (1) *which parameters are relevant for the algorithm's performance* and (2) *what is the best value for each important parameter?*

The text is structured as follows. In the section 2 the expressions to predict the number of I/O operations and CPU performance are introduced. The section 3 explains the system architecture of the software implemented to carry out the tests using real and synthetic data sets. The set of rules are described and justified in the fourth section. Section 5 presents some conclusions and suggestions.

2 Spatial Join Algorithms

The algorithms that perform the filtering step of the spatial join operation manipulate object descriptors, defined by their object identifier (OID), the minimum bounding rectangle (MBR) and a pointer to the geometric description of the object.

Figure 1 shows a complete set of algorithms that perform the filtering step of a spatial join operation. The algorithms are classified according to the file organization used to maintain object descriptors. As mentioned before, we select a representative algorithm for each class, except for the nested loops and one-indexed file classes. In preliminary tests, nested loops presented very long response time, being the worst choice in any situation. The algorithms in the class "one indexed file" are an extension of the "pure" algorithms for a special case, and we expected that the results can be extended for them.

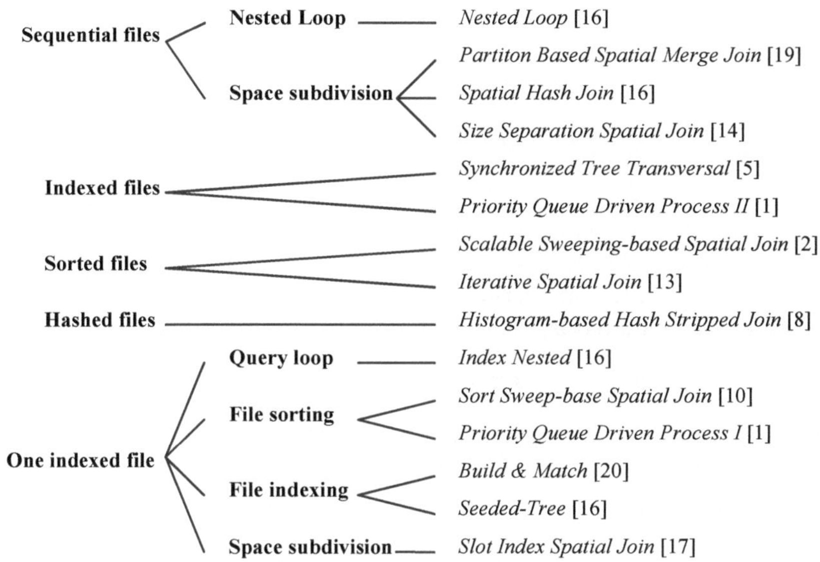

Fig. 1. Classification of spatial join filter step algorithms according to their method of file organization.

2.1 Plane-sweep technique

All spatial join algorithms, in different ways, load objects into memory and a kind of plane-sweep strategy to check if object pairs satisfy the spatial predicate[6, 21].

The plane-sweep technique has a performance of $O(k + n \log n)$, where k represents the number of intersections between objects and n is the number of objects [6]. If the technique is used to find intersections between objects of two sets, then $n = n_A + n_B$, where n_A and n_B is, respectively, the number of objects in sets A and B. The value of k can vary a lot, depending on the spatial distribution and the size of objects.

In order to predict the algorithm performance in the most accurate way, we define the number of pairs for which the spatial predicated is verified, expressed by c, independently if the result is true or false. Figure 2 shows an example where the number of intersections (k) is zero, but the number of comparisons (c) is six, because the pairs (o_2,o_1), (o_3,o_2), (o_3,o_1), (o_4,o_3), (o_4,o_2) and (o_4,o_1) are tested.

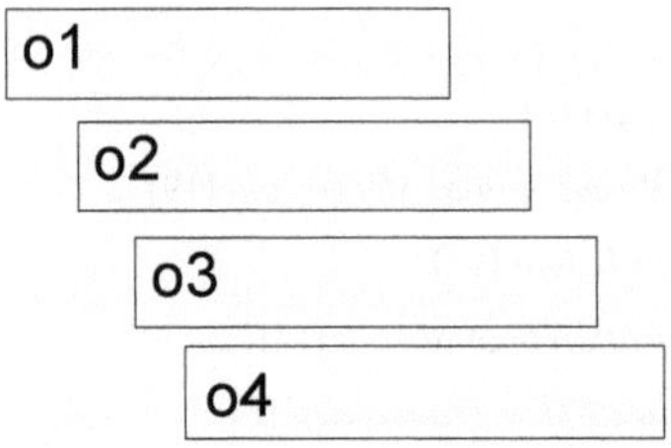

Fig 2. An example of a set of objects where $k=0$ and $c=6$.

In a first approach, considering the objects uniformly distributed in the space, c can be estimated by

$$c = s_x n \tag{1}$$

where s_x, represent the average size of objects. This technique also works if the objects are sorted by their y coordinates. The estimation in this case is $c=s_y n$.

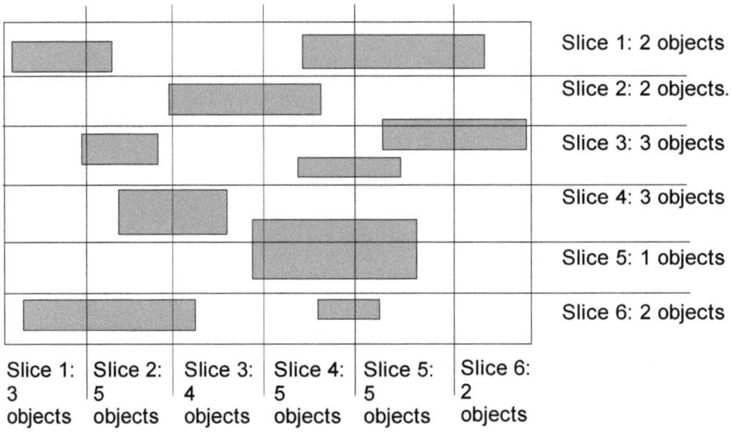

Fig. 3. An example of space stripping.

For non-uniform distributions, we divide the space into strips, as can be seen in figure 3. For each strip, the number of overlapping objects is counted, and a object distribution histogram is built [3,4]. The estimation can be made by the following expression, where Φ is the number of strips and n_i is the number of objects of sets A and B in strip i, for $1 \le i \le \Phi$.

$$c = \sum_{i=0}^{\Phi} n_i \tag{2}$$

Although the size of objects might not appear to be important, it has an impact in the number of objects in each strip. Figure 3 shows an example of a set of objects in a space divided by 6 strips. As can be observed, the shape of objects significantly modifies the number of objects per slice.

The plane-sweep technique is adapted according to the spatial join algorithm. These modifications can alter the value of c, and are described within the respective algorithm.

2.2 Synchronized Tree Transversal

The Synchronized Tree Transversal (STT) [5] algorithm needs that both sets are indexed, in advance, for two different R-Trees, named R_A and R_B. Nodes of both R-Trees are compared, in a synchronized way. The number of nodes comparisons was defined in [11], as

$$Comp = \sum_{i=1}^{h-1} \sum_{1}^{n_i^A} \sum_{1}^{n_i^B} (\min(s_{x,i}^A + s_{x,i}^B),1) \times (\min(s_{y,i}^A + s_{y,i}^A),1) \tag{3}$$

Initially, the height of both trees are considered to be equal and is expressed just by h. Be n_i^A the number of nodes of set A in level i of the R-Tree R_A, and $1 \le i < h$. The average size, in the axis x, of nodes or MBR in level i is represented by $s_{x,i}^A$.

Based on [11, 22, 23], the number of I/O operations when the buffer size is equal to zero is defined as $Zj = 2 + 2Comp$. Introducing a buffer, ρ represents the possibility of a certain node being found in buffer memory, reducing the number of I/O operations to

$$io_{STT} = n_R^A + n_R^B + (Zj - n_R^A - n_R^B)\rho \tag{4}$$

The expression for the CPU performance depends on the number of nodes comparisons (3). For each node comparison, *2Fanout* objects are involved, at maximum, resulting in

$$cpu_{STT} = O(c_{STT} + 2Fanout \log_2 Fanout) \tag{5}$$

An adequate value for c_{STT} is necessary to complete the expression. The R-Tree divides the space in an irregular way. Two objects, aligned on the x axis, can be allocated in different nodes. As a result, they will not be compared to evaluate the spatial predicate, reducing c. Let s_1^R be the average size of leaves, c can be estimated by

$$c_{STT,1} = \frac{c}{s_1^R} \tag{6}$$

The value of c, for internal levels, depends on the number and average size of nodes of the lower level.

$$c_{STT,i} = \min((s_{x,i+1}^A + s_{x,i+1}^B),1) \times n_{i+1}^A \times n_{i+1}^B \tag{7}$$

Add the number of comparisons in all levels, the value of c for the entire R-Tree is obtained.

$$c_{STT} = \sum_{i=1}^{h} c_{STT,i} \tag{8}$$

2.3 Iterative Stripped Spatial Join

The Iterative Stripped Spatial Join (ISSJ) algorithm is an adaptation of the Iterative Spatial Join, proposed by Jacox and Samet [13]. The main idea is, first, to sort the sets of objects separately. Then, the plane-sweep technique is applied, scanning both sets in sequence. During the plane-sweep, just the active objects are maintained in memory. The space is divided in strips to reduce the number of comparisons. When an object is loaded to memory, it is assigned to one or more strips. A different active objects list is maintained for each strip. In this way, pairs of distant objects are not compared.

Considering a uniform distribution of objects, in each strip the number of objects is $r(n^A + n^B)/\Phi$, where r represents the replication factor. It can be calculated as the number of objects, with replicas, over the original number of objects, $r = (n_R^A + n_R^B)/(n^A + n^B)$, resulting in a value greater than 1.

The number of I/O operations depends on whether the sorting is internal or external. The best case occurs if both sets are sorted using an internal algorithm. Then, the number of I/O operations is

$$io_{ISSJ} = 3(b^A + b^B) \tag{9}$$

The worst case occurs if both sets are sorted by an external algorithm. The number of I/O operations increases to

$$io_{ISSJ} = (b^A + b^B) + 2b^A \times \left| \log_{M_b} b^A \right| + 2b^B \times \left| \log_{M_b} b^B \right| \quad (10)$$

Due to the space division in the strips, the number of pair comparisons is reduced to

$$c_{ISSJ} = \frac{rc}{\Phi} \quad (11)$$

The expected performance of the algorithm is

$$cpu_{ISSJ} = O(c_{ISSJ} + r(n^A + n^B)\log\frac{r(n^A + n^B)}{\Phi}) \quad (12)$$

2.4 Partition Based Spatial Method

As a first step, the PBSM [19] divides the space into a set of cells by applying a regular grid to it. A partition is a set of cells and each cell belongs to one, and only one, partition. A spatial object may intersect one or more adjacent cells, belonging to different partitions. The algorithm replicates the object descriptor in all intersecting partitions.

The objects are loaded to memory, distributed in partitions, and then reloaded to memory to the plane-sweep. For each object, three I/O operations are executed. But, a certain percentage of partitions, represented by o, overflows, because all objects don't fit in memory at the same time, forcing another read/write operation. Replicas in the result set are avoided using the *Reference Point Method* (RPM) [7]. Thus, the number of I/O operations can be expressed by:

$$io_{PBSM} = (2or^2 + 2r + 1)(b^A + b^B) \quad (13)$$

Due to space subdivision, the number of pairs of objects comparisons is reduced. If the number of horizontal cells is greater than the number of partitions, as figure 4a shows, in a same row, more than one cell can be allocated to the same partition. If the number of horizontal cells is smaller, as in figure 4.b, considering just one row, each cell is allocated to a different partition. Being the number of horizontal cells represented by Φ_H, the value for c can be estimated by

$$c_{PBSM} = \begin{cases} \dfrac{c}{\Phi_H}, P > \Phi_H \\ \dfrac{c}{P}, P \leq \Phi_H \end{cases} \quad (14)$$

1	5	4	3
2	1	5	4
3	2	1	5
4	3	2	1

(a)

1	2	3	1
2	3	1	2
3	1	2	3
1	2	3	1

(b)

Fig. 4. Cell allocation, according to the number of partitions, using a round-robin schema.

In a uniform distribution, all partitions have the same number of objects, including replicas. In this case, the plane-sweep technique for each partition processes $r(n^A + n^B)/P$ objects. As the plane-sweep is repeated for all partitions, the total number of comparisons is expected to be in an order of

$$cpu_{PBSM} = O(c_{PBSM} + r(n^A + n^B)\log\frac{r(n^A + n^B)}{P}) \quad (15)$$

2.5 Histogram-based Hash Stripped Join

The Histogram-based Hash Stripped Join (HHSJ) [8] has three main characteristics: the object descriptors are stored in a hash file organization; a bi dimensional histogram of object distribution defines the spatial extension of each partition; and, when it loads objects to memory, it divides the space by strips.

The histogram is maintained in a quadtree, so-called *HistQ*. Each of its quadrants represents a subdivision (a cell) of the global space. Each node contains a counter of the number of objects that intersects the space segment represented by this node of the quadtree. Each level of the quadtree represents a different histogram of object distribution, with a different degree of precision.

A hash file is created for each data set. It is organized in buckets, assuming a static hash organization. The hash function associates a leaf node of *HistQ* to a specific bucket in the hash file. Therefore, there exists one bucket in the hash file for each leaf of the quadtree. If an object transposes quadrants and is counted in two (or more) leaf nodes of *HistQ*, its descriptor is replicated in two (or more) buckets. The number of replicated objects is defined during the creation of the hash file.

The minimum number of buckets is defined by

$$Buckets = \left\lceil \frac{n}{Fanout} \right\rceil \tag{16}$$

Using (16), the height of the *HistQ* can be calculated by

$$h = \lceil \log_4 Buckets \rceil \tag{17}$$

In fact, the number of buckets, and, considering a disk block for each bucket, is

$$b = h^4 \tag{18}$$

Because some buckets in the hash file can overflow, some additional disk blocks are necessary.

To carry out the test of spatial predicate in pairs of objects, first, the HHSJ loads the *HistQ* of both sets to main memory. Based on them, partitions and respective boundaries are defined. Each partition is spatially defined by the corresponding selected quadrants in both quadtrees. The space is divided in an irregular way. Regions more densely populated are divided into small areas, to maintain the number of objects pertaining to a partition into the available memory.

Before, the algorithm processes each partition separately. First it identifies the buckets that are covered by the partition. Then, the object descriptors stored in each bucket are loaded to memory.

The algorithm just loads to memory both *HistQs* and all buckets, including overflow buckets, of each hash file. So, the number of I/O operations is

$$io_{HHSJ} = b_A^{HistQ} + 4^{h_A} + b_A^{Over_{HistQ}} + b_B^{HistQ} + 4^{h_B} + b_B^{Over_{HistQ}} \tag{19}$$

The number of comparisons between pairs of objects is reduced by two factors: the number of partitions (*P*) and the number of strips inside each partition (*l*). As the space is divided in an irregular way, we use the mean partition size, expressed by s_x^{Part} and a constant number of strips, for all partitions, resulting in

$$c_{HHSJ} = \frac{c}{s_x^{Part} \times \Phi} \tag{20}$$

The performance in CPU can be estimated by

$$cpu_{HHSJ} = O(c_{HHSJ} + r(n^A + n^B)\log\frac{r(n^A + n^B)}{P\Phi}) \tag{21}$$

3 The System Architecture

The performance analysis, although correct, simplifies many cases. To compare the algorithms in real situations, a software system was implemented to carry out the tests between them.

To acquire real data sets, one tool converts different data formats to the internal files. Another tool generates synthetic data sets. This second type of data is valuable for performance tests, because the user can control the parameters that describe the data set characteristics, changing one or two of them in each data set, and then test the algorithms to verify differences in the their performance.

The main component in the system architecture is the join module, which contains the spatial join algorithms implemented. The design of the join module allows one to plug and play other algorithms easily.

When running a test, the user specifies the desired algorithm and the size of main memory. The buffer capacity is measured in number of pages and each page has a fixed size of 4Kb. For the STT algorithm, a specific R-Tree Oriented Buffer was implemented for better performance. The other three algorithms use a traditional LRU buffer.

For all algorithms, the overall response time and number of I/O operations can be obtained. For specific algorithms, other values, like the replication factor or the size and height of R*-Tree, can be obtained.

The entire system was implemented in C language, for Linux operating system, by us. During tests, the system operating buffer was turned off, so as to not influence the result time. All tests were performed on an Intel 2.4GHz, with 512 Mb of RAM and SCSI disks.

Table 1. Real Data sets Characteristics

Name	Description	Cardinality	# of disk blocks
Source: R-Tree Portal (www.rtreeportal.org)			
ca_streets	Californian streets - multilines	2.249.727	13.157
ca_streams	Californian streams of water - multilines	98.451	576
ge_roads	German roads - multilines	30.674	180
ge_utility	German public utility networks - multilines	17.791	103
ge_rrlines	German railroads - multilines	36.334	213
ge_hypsogr	German hypsographic data - multilines	76.999	451
gr_roads	Greek roads - multilines	23.278	137
gr_rivers	Greek rivers - multilines	24.650	145
la_streets	Los Angeles streets - multilines	131.461	769
la_rr	Los Angeles railroads - multilines	128.971	755
Source:Bureau of Transportation Statistics (USA) – www.bts.gov			
usa_counties	Counties - polygons	3.236	19
usa_rr	Railroads - multilines	166.688	981
usa_hydro	Rivers - multilines	517.538	2959

The artificial data sets were generated by the system, and the real data sets were obtained from different sources. Table 1 shows the name, description, cardinality and average size on both axes for each real data set, grouped by its source. As can be seen, we used a great number of real data sets, with very different characteristics. The cardinality has an obvious importance, because it defines the workload for the algorithm. The average size of objects is important because it defines, to a great extent, the number of pairs of objects that will be compared.

4 Rules for Performance Optimization

In this section, the set of rules for performance optimization is presented. Each rule is explained and exemplified, showing the obtained results.

Rule 1: the DBMS can estimate *k* for each axle and choose the one with minor value of *k,* optimizing the plane-sweep.

The rule proposes that the GDBMS, first, must estimate the number of comparisons to be made in each axle, and, after, performed the spatial join algorithm sorting the data by the chosen axle. Two estimation approaches are suggested, one based on equation (1), another based on histograms (2).

Firstly, we generate a number of synthetic data sets, varying the shape of the MBR, from a relation of 5:1 to a relation of 1:5. The cardinality of data sets are, always, 200.000 objects. The graph in figure 5 shows the response time for each algorithm, sorting by the same axle in all tests, making clear the minimizing possibilities of the axle choice.

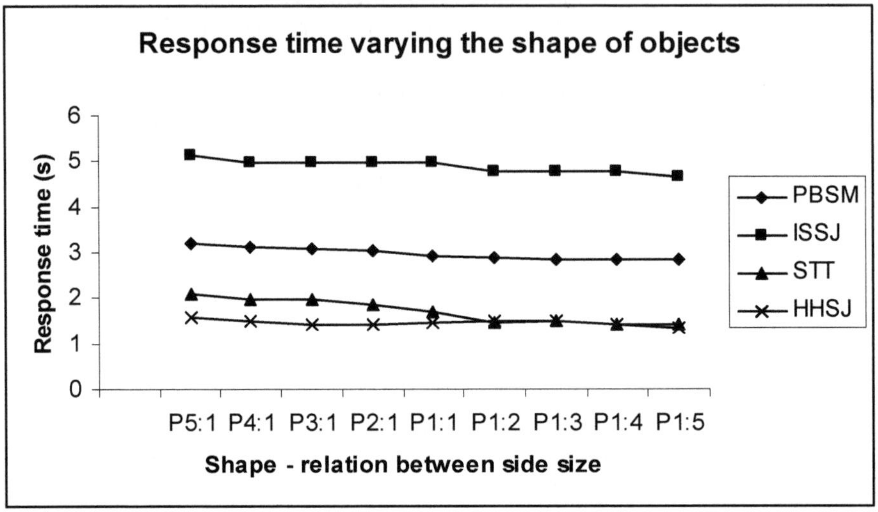

Fig. 5. Graph of response time varying the shape of objects.

We run all possible joins between real sets, sorting by both axles to validate both alternatives. When the difference between response time was less than 5%, we discard the join, because the difference is closed that both sorting alternatives seems to be good enough. In this way, we concentrate in greater differences, where the choice is relevant. Alternative 1, based just in mean size, chooses the right axle in 29 times, but the wrong axle 10 times, because it is a rough simplification of the data sets, representing a 74.3% of correct match. Alternative 2 presented a 100% of correct match, when the space is divided in 500 strips. The number of correct match is not affected by the spatial join algorithm.

This result indicates clearly that the histogram based prediction is superior and must be used always that is possible, justifying the necessity to maintain the histograms always that an object is inserted or deleted from the data set. But, in a query plan, the spatial join can be performed after another operation, like a selection. In this case, the histograms are not available. Our suggestion is, during the anterior operation, creates the histogram, according to the temporary set is written.

Rule 2: The STT algorithm is optimized defining nodes with a low number of entries. The total number of nodes will be greater, elevating the value of *Comp*, and defining a minimum limit for the rule.

The rule 2 intends to optimize the STT algorithm. The construction of R*-Trees is performed using the STR algorithm [15] to reduce the number of nodes and obtain a maximum occupation of entries in each node. This method reduces the number of I/O operations, because the size of R*-Trees are minimized. Second, the buffer algorithm try to maintain non-leaf nodes in the memory buffer, because non-leaf nodes are more susceptible to reloads than leaf nodes, also, reducing the number of I/O operations.

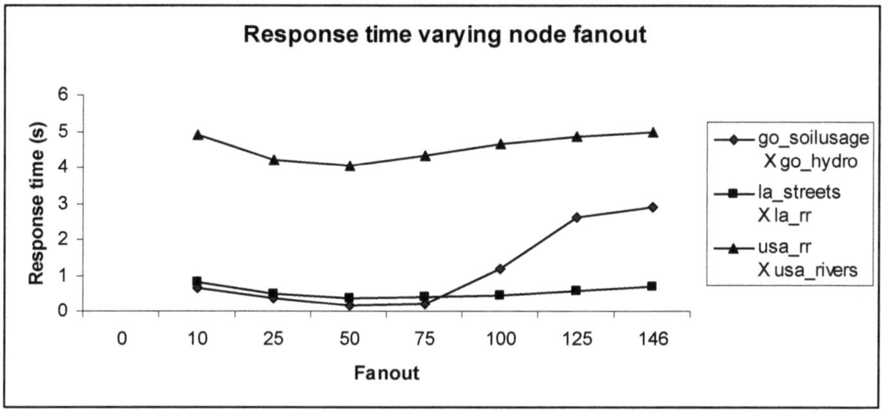

Fig. 6. Graph of response time using the STT algorithm, varying the fanout.

The rule changes the number of entries in a node, called *fanout*. The number of leaf nodes can be calculated as $n_{Leafs} = \lceil n / Fanout \rceil$. The number of non-leaf nodes, is proportional to the number of leaf nodes. So, if the fanout is reduced, the number of nodes are incremented, maybe, adding a new level to the R*-Tree. By expression (3), the number of node comparisons is, also, increased, limiting the rule 2. Figure 6 shows the response time for three different cases of spatial join, varying the fanout from 10 to 146, which is the maximum for a node written in just one block.

Also, the influence of the buffer memory size was verified. The performance of STT algorithm is constant when the memory size increases, confirming the results showed by [9]. In fact, the time to execute I/O operations represents a small part, in general, less than 2% of the total time, although the buffer hit ratio increases according to the memory size, reducing the number of I/O operation performed, as expected.

Rule 3: The ISSJ algorithm is optimized defining a great number of strips. The number of objects in each strip will be small, but the rule is limited by the adding of replicas.

The third rule is to increase the number of strips to optimize the algorithm ISSJ. But, this rule increments the replication factor, which penalizes the performance.

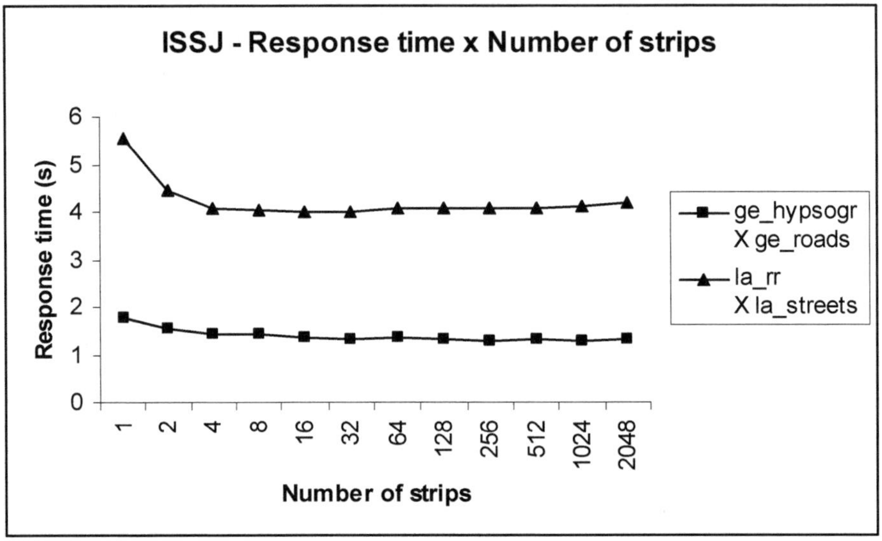

Fig. 7. Graph of response time and replication factor, for the ISSJ, varying the number of strips.

The graph in figure 7 shows the response time of three spatial joins, varying the number of strips and the replication factor. As can be seen, a small number of strips, four to eight, present an increment in algorithm performance, reducing the response time. After this number of strips, the response time almost stabilizes or presents a small increment, because the number of processed objects (original + replicas) increases exponentially.

Changing the available memory size, the performance of the algorithm suffers great impact of changing the sorting algorithm. The number of I/O operations is reduced, as well as, CPU cost. In the graph of figure 8, this fact is clear, with a step down according to the increase of memory size, in the exact point where there is enough memory to do an in-memory sorting of one of the joined sets. All operations were performed with a constant number of 8 strips.

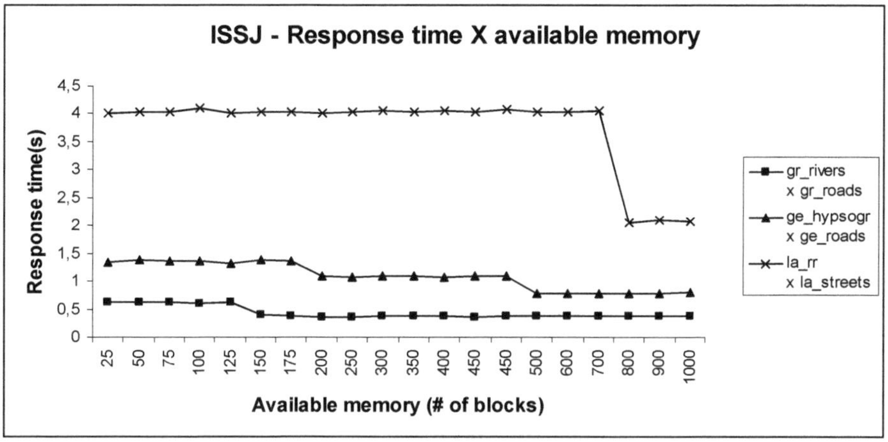

Fig. 8. Graph of response time for algorithm ISSJ, varying the available memory size.

Rule 4: The PBSM algorithm is improved setting a high value for the number of partitions using a small size of memory or just set a lower bound to the number of partitions. This rule is limited by the number of replicas, which increase the number of processed objects.

The rule 4 increases the performance of the PBSM algorithm by incrementing the number of partitions. In this rule, the replication factor is, also, increased, impacting the performance. The GDBMS can allocate less memory to the algorithm to control the number of partitions or directly set a value for *P*. Figure 9 shows the graph of response time of three spatial join operations, for which the available memory size was changed, to induce different number of partitions. As the memory size increases, the number of partitions is reduced, in general, increasing the response time, as expected from (15). In all cases, the algorithm uses a 32x32 grid of cells.

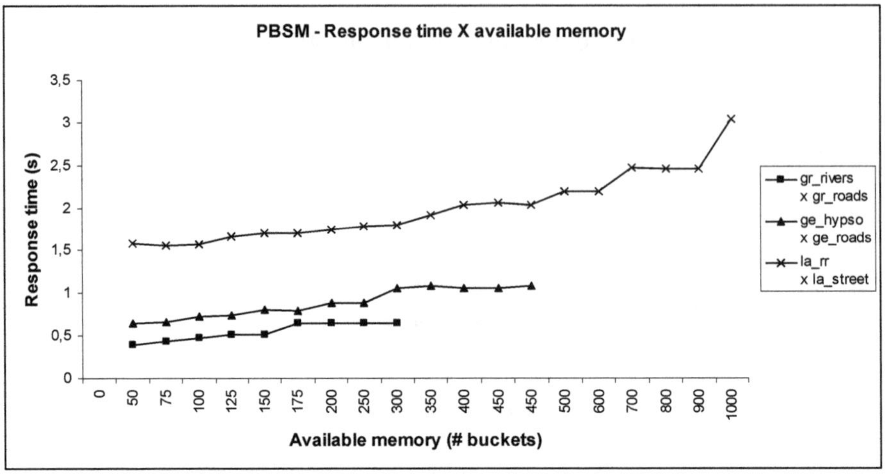

Fig. 9. Graph of response time for the algorithm PBSM, varying the available memory size.

Rule 5: The HHSJ is improved setting a large value for the number of partitions and for the number of strips. This rule is limited, also, by the number of replicas, which increase the number of processed objects.

The rule 5 is a combination of the strategies 3 and 4, as the HHSJ algorithm combines aspects of both the ISSJ and the PBSM algorithms. Replication can occur in two moments when HHSJ is executed: during the creation of the hash file; and when objects are loaded into memory and divided by strips. As the creation of the hash file can be done only once when the set is loaded to the GDBMS, we concentrate our attention in the second moment, which occurs always when the set is involved in a spatial join operation.

Figure 10 shows a graph of response time against available memory size. The number of partitions varies according to memory size, but we kept in 16 the number of strips in all cases. For small sets, the response time is almost constant. For larger sets, the response time increases with the available memory size, being constant after a certain amount of memory. This behavior is different than PBSM, where the response time just increase, not stabilizing. In HSSJ, the creation of strips, in memory, allows the stabilization, establishing a maximum value for the response time. For the largest set spatial join operation, *ca_street* and *ca_stream,* the stabilizing point is achieved just at almost 8Mb of memory, but it is not visible in the graph.

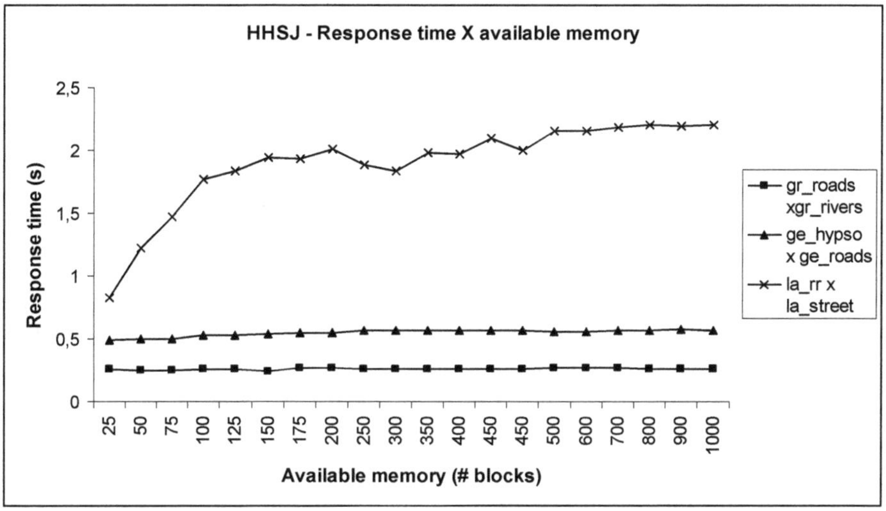

Fig. 10. Graph of response time for the algorithm HHSJ, varying the available memory size.

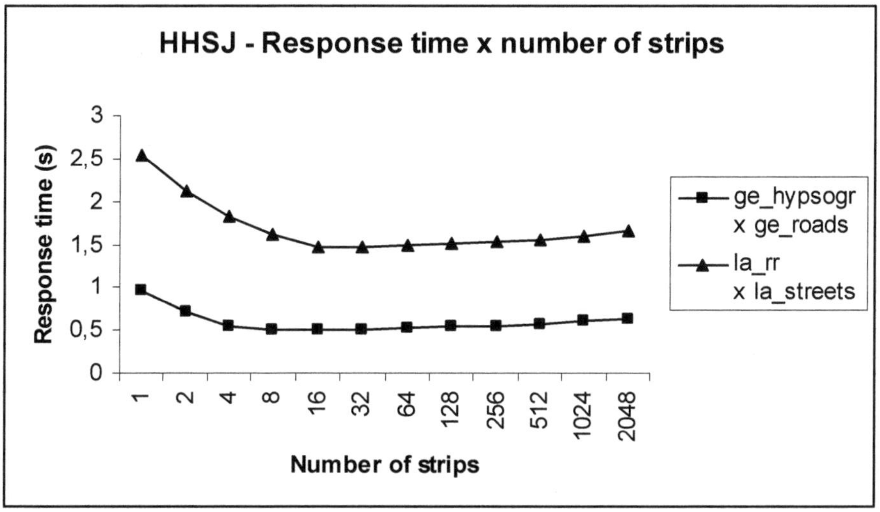

Fig. 11. Graph of response time and replication factor for the algorithm HHSJ, varying the number of strips.

The graph in figure 11 shows the effect of change the number of strips in algorithm HHSJ, maintaining the available memory size. The behavior of response time is almost the same of the ISSJ: a small number of strips results in an important decrease of response time; a great number of strips produces many replicas, increasing the response time.

If the available memory size increases, the number of I/O operations is not changed, because the algorithm read all buckets only once to perform the operation.

5 Conclusions

This paper first introduces expressions to predict the number of I/O operations and CPU performance of each studied algorithm, using an unified notation. After, a set of simple rules are established to improve the performance of such algorithms. Using tests with real and synthetic, response times were collected to prove the correctness of rules. In some situations, for example, spatial joining the sets *ca_streets* and *ca_stream* using PBSM, the response set varies between 20.3s and 48s, a difference of more than 50%. In some extreme cases, like optimizing the fanout in R-Trees, the STT algorithms can perform the spatial join more than 10 times faster.

The study concentrates in spatial join operation using bi-dimensional sets, but the proposed strategies can be easily extended to operate in three or more dimensional spaces and spatiotemporal data structures. Although tests with real data sets permit the evaluation of very different scenarios, we do not expect to cover all possibilities. The experiments can be carried out for other system architectures like palmtops, where some additional constraints can impact the performance. Another possibility is running in a grid architecture, but this imply a significative change in the code of investigated algorithms.

The goal of reducing CPU time can be explored in many other areas, like statistic databases, clustering spatial objects and text mining algorithms. Another aspect to investigate is the implications of CPU caches with 2Mb or more, in traditional databases algorithms.

Also, we propose that available GDBMS, like Oracle Spatial [18], IBM DB2 Spatial Extender [12] and PostGis [20] can incorporate the proposed rules and more alternative file organizations, like sorted and hash files, and algorithms, like ISSJ and HHSJ. In general, they implement only some variation of R*-Trees, reducing the number of spatial join algorithms to just three: STT, if both sets are spatial indexed; Scan&Index, if one set is indexed; and nested loops, if there is no indexes. But, even for these algorithms, some rules can increase the performance when executing spatial joins, because the differences in response time are large.

References

[1] Arge, L., O. Procopiuc, O. et al. (2000) "A Unified Approach for Indexed and Non-Indexed Spatial Joins". Proc. of 7th Int'l. Conf. on Extending Database Technology, p. 413-429, 2000.

[2] Arge, L., Procopiuc, O. et al. (1998) "Scalable Sweeping-Based Spatial Join" Proc. of VLDB, 1998, p.570—581, 1998.

[3] Belussi, A.; Bertino, E. and Nucita, A. (2003), "Grid Based Methods for Spatial Join Estimation" Proc. of 11TH Symp. of Advanced Databases Systems, p.49-60, 2003.

[4] Belussi, A.; Bertino, E. and Nucita, A.(2004). "Grid Based Methods for Estimating Spatial Join Selectivity" Proc. of 12th ACM GIS, 2004.

[5] Brinkhoff, T.; Kriegel, H.P. and Seeger, B. (1993), "Efficient processing of spatial joins using R-trees" Proc. of ACM SIGMOD, p. 237-246, 1993.

[6] De Berg, M, M. Van Kreveld, M. Overmars and O. Schwarzkopf (2000), "Computational Geometry", 2nd. edition. Springer-Verlag, 2000.

[7] Dittrich, J.P. and Seeger, B. (2000), "Data Redundancy and Duplicate Detection in Spatial Join Processing" Proc. of Int'l. Conf. on Data Engineering, p. 535-543, 2000.

[8] Fornari, M.R. and Iochpe, C. A spatial hash Join algorithm suited for small Buffer Sizes. Proc. of 12th ACM GIS, p. 118-126, 2004.

[9] Gatti, S.D. and Magalhães, G.C. (2000), "A Comparison Among Different Syncronized Tree Transversal Algorithms for Spatial Joins" Proc. of GeoInfo, 2000 - available in www.geoinfo.info.

[10] Gurret, C. and Rigaux, P. (2000) "The Sort/Sweep Algorithm: A New Method for R-Tree Based Spatial Joins" Proc. of Statistical and Scientific Database Management, p.153-165, 2000.

[11] Huang, Y.W. and Jing, N. (1997), "A Cost Model for Estimating the Performance of Spatial Joins Using R-Trees" Proc. of Int'l Conf. on Information and Knowledge Management, 1997.

[12] IBM (2005) , "IBM DB2 Spatial Extender- User's Guide and Reference Version 8", 2005.

[13] Jacox, E.H. and Samet, H. (2003), "Iterative Spatial Join" ACM Transactions Database Systems, v.28, n.3, p. 230—256, 2003.

[14] Koudas, N. and Sevcik, K.C.. (1997) "Size Separation Spatial Join", Proc. of ACM SIGMOD, p. 324-335, 1997.

[15] Leutenegger, S. T.; Edgington, J. M.; Lopez, M.A. (1997) "STR: A Simple and Efficient Algorithm for R-Tree Packing". Technical Report- TR-97-14. University of Denver, Mathematical and Computer Science Department, Denver: USA.

[16] Lo, M-L. and C.V. Ravishankar (1996), "Spatial hash-joins" Proc. of ACM SIGMOD, 1996, p. 247—258, 1996.

[17] Mamoulis, N. and Papadias, D. (2003), "Slot Index Spatial Join" IEEE Trans. on Knowledge and Data Engineering, v.15, n.1, p. 211-231, 2003.

[18] ORACLE, "Oracle Spatial User´s Guide and Reference 10g Release 1", Dec., 2003.
[19] Patel, J.M. and DeWitt, D.J. (1996), "Partition Based Spatial-Merge Join" Proc. of ACM SIGMOD, p. 259-270, 1996
[20] REFRACTIONS RESEARCH (2005), "PostGIS Manual." Disponível em http://postgis.refractions.net/docs/postgis.pdf.
[21] Rigaux, P.; Scholl, M. and Voisard, A. (2000), "Spatial Databases with Applications to GIS". Morgan Kaufmann Pub., 2000.
[22] Theodoridis, Y.; Sellis, T. (1996) "A Model for the Prediction of R-Tree Performance". Proc. of 16th. ACM Symposium on Principles of Database Systems, ACM Press, 1996, p.161-171.
[23] Theodoridis, Y.; Stefanakis, E.; Sellis, T. "An Efficient Cost Model for Spatial Joins Using R-Trees" IEEE Transactions On Knowledge And Data Engineering,v.12, n.1 , p.19-32, January,2000.

Spatial Query Broker in a Grid Environment

Wladimir S. Meyer, Milton R. Ramirez, Jano M. Souza

Instituto Alberto Luiz Coimbra de Pós Graduação e Pesquisa de Engenharia
Universidade Federal do Rio de Janeiro

1 Introduction

Lately, the profusion of equipment related to spatial data generation has been responsible for the production of a lot of spatial data to attend different purposes in many domains. Government agencies, large private corporations and scientific centers are the most common producers and consumers of this kind of data that are employed in decision-making systems, analysis tools and experiments.

Many efforts have been made to create standard architectures capable of offering several levels of relationship among spatial data producers and consumers. The Open Geospatial Consortium (OGC) is one of the most important organizations involved with the standardization and its interfaces and services have been recommended as a solution for interoperability problems over a distributed environment. Even though OGC standards have been adopted by many systems, as a solution for uniform spatial data access, there is a lack of mechanisms capable of leading them towards the high performance direction, as in services related to data transfer and spatial data processing among servers [1]. A typical question is "how to dynamically execute spatial operations involving features from different data providers?" In these situations the client application is responsible for taking care of the entire process after accessing the remote spatial data. So all complexities must be embedded in the client application, which doesn't necessarily lead to an efficient result.

The grid-computing paradigm gathers many capabilities that could be employed in the aforementioned context. It has emerged with the purpose of sharing free resources beyond organizations' boundaries, aimed at a high performance infrastructure.

This work aims to provide a strategy to execute spatial joins among large spatial databases spread geographically and integrated by a grid environment. This strategy is implemented as a spatial query broker being capable of receiving a spatial query and selecting computer nodes dynamically to process parts of the original query in parallel in order to reduce its overall execution time. The architecture was influenced by previous ones presented in [2], [3] and [4].

The structure of the paper consists of an explanation of problems related to distributed spatial queries (section 2) and some solutions based on filter/refine strategy, since they play an important role in the GIS domain, followed by an overview of Virtual Organizations in a grid context (section 3) as a metaphor, emphasizing their aspects and functionalities that could be applied to a distributed GIS context. A description of resource brokers (section 4) means to depict important functionalities that could be used in a spatial query broker. Some related works involving distributed spatial query processing are presented in section 5 and the proposed architecture is detailed in section 6. Some preliminary results on the parallel execution of spatial queries are analyzed in section 7 and some important remarks are taken into account in section 8.

2 Distributed Spatial Queries

A filter/refine technique has been adopted in many spatial database management systems (SDBMS) to reduce the exact geometric tests in spatial queries [5]. It consists of the execution of a preliminary query with approximations of the actual features (usually Minimum Bounding Rectangle - MBR) to discard some of those that don't satisfy a specific. The next step consists of executing the exact processing of the involved geometries to determine which of them definitely satisfy the predicates.

An important characteristic of spatial queries is that they are much more expensive than those involving conventional data. The size and number of vertices of geometrics' data are the main factors. The queries normally use large computing resources when executing operations with geometrics and/or topologic predicates, turning indispensable the use of spatial indexes.

This technique was extended in [6] to process spatial joins, including a geometric filter step after a MBR join. This new step introduces a more accurate approximation of spatial objects, to reduce the size of inconclusive pairs.

Some strategies were proposed in [7], based on Brinkhoff's model to deal with spatial joins in a distributed environment trying to explore an additional parallelism among SDBMS involved during the exact processing step.

Trying to improve response time a proposal is presented in [8] where the authors aim at executing the two steps of the filter/refine technique in parallel, but details about preliminary considerations limit the scope of the proposal: high bandwidth and thematic data partitioning.

In spite of the good results presented by previous solutions, a small number of servers is involved during the exact geometry processing, normally those storing the themes, which can lead to inadequate response time when dealing with very large data sets.

3 Virtual Organizations in a Grid context

When organizations, computational resources, services and people with similar interests are put together and the sharing rules is defined, they originate a so called *Virtual Organization (VO)* [9]. The VO is an abstraction that can involve many actual organizations interested in collaborative work, i.e., they can share services, data and computational resources to reach some common goal. This abstraction could be adopted, for example, by organizations interested in sharing spatial databases or making up a distributed GIS.

A VO often presents some basic functionality items that may be accessed by any member nodes. Some of the most important are:

- File catalog – It consists of a service specialized in registering and monitoring file replicas spread among nodes and that may have distinct logic names;
- Job manager – Entity that receives jobs execution requests and verifies all necessary resources for their execution;
- Resource broker – A service capable of checking all necessary resources described in a job specification and invoke all those resources to perform its execution;
- Information service – A suite of services that gathers information about resources and are often used by applications and other services (like the resource broker).

The profile of a VO can change as a result of its purpose. Services, resources and topology may assume different configurations depending on the nodes' capabilities, the nature of the tasks to be performed and the interests of the members.

Despite the flexibility and the computing-on-demand offered by the VO abstraction, some difficulties like heterogeneity and security, are great challenges for the research carried out [10,11].

The adoption of standards like OGSI (Open Grid Service Infrastructure) and, more recently, WSRF (Web Service Resource Framework) has brought a new perspective to the VO environment since now it can deliver services with a high degree of standardization for the use of Web services technology. Security problems related to the data and binary transmission may be avoided with the use of binaries encapsulated into services that offer only a standard interface, preserving code detail. In a similar way, the heterogeneity problem inherent to a grid environment can benefit from these standard interfaces since the details related to service implementation are omitted from the consumers of the resources, improving both integration and scalability as a whole.

So, as depicted above, a VO takes an important role in a grid environment and should provide many high level services in order to permit the increase of collaboration among its members. Resource brokers take an important place in this context, especially when the application's nature involves hundreds or thousands processors nodes and resources.

4 Resource Brokers (RB)

In an environment, with thousands of resources dynamically changing, it is almost impossible for a user to choose the best resource to execute a job without any previous information. The RB acts as a middle tier between users and resources [12] and assumes the responsibility of finding the best resources for them and, complementarily, passing the job specification to the resources found (Figure 1).

In this figure, the GIS module means "Grid Information Service", the component responsible for advice on resources available. The resource broker - RB - uses this information to locate the best resources without explicit users' participation.

Resource brokers are not present natively in all grid middleware, but many initiatives have generated products that turn the grid user's life easy. GridBus [13], Emperor [14] and GridWay [15] are some of these efforts.

They can be thought as high level services that locate resources that agree with a job specification.

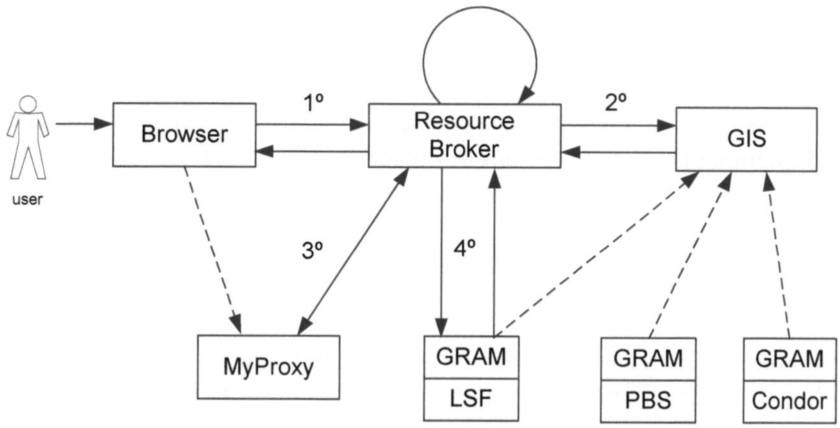

Fig. 1. RB acting as a middle tier in a grid - adapted from [12]

The role of scheduler may also be taken up by the RB, who makes execution plans and defines the order in which tasks are to be carried out based on complementary information like resource capabilities, communication costs and other statistic metrics. For this ability, RBs are also known in grid platforms as "meta-schedulers".

When assuming a scheduler's role, problems concerning the multi-organizational nature of the grid take place: difficulty to pre-allocate, load balancing and different performance for communication channels, besides the natural resources' heterogeneity [11,9].

There are two basic strategies adopted by resource brokers to create an execution plan when acting as a scheduler: static and dynamic strategies. The first one defines the execution plan before runtime and any context changes occurred during runtime are not taken into account; an example of this class of RB is the WMS (Workload Management System) [16] which integrates the gLite grid middleware. The second one starts with a predefined plan that can change during runtime, in an adaptive manner, chasing the best results as a whole. GridWay RB [15] lies in this class.

Scientific applications, the greatest users and motivators of the grid paradigm, are strongly based on file as the lowest data unit [17]. This implies that most of the RBs are tailored to this data approach.

This paper proposes a similar service to deal with scheduling of spatial queries among spatial databases systems in a grid, similar to the job/file approach of conventional RB. The *Spatial Query Broker (SQB)* being proposed is presented in section 6.

5 Related Works

During last years many efforts have been made to permit executions of spatial queries in a distributed environment.

An extension of Brinkhoff solution, for the distributed spatial query processing was proposed by [7] with the purpose of exploring the parallelism in the most critical phase of a spatial join process: exact geometry processing. In that work, after the elimination of the geometries that definitely don't participate from the final result, by means of approximate filtering, the inconclusive geometries are simultaneously processed by the SDBMS that participate of the query. This strategy was called MR2 and uses spatial indexes (R*-Tree), from themes involved in the operation, during the filtering step. In this proposal only execution of a query is taken into account. This adaptation of the three-step approach is followed by the present proposal since the filtering phase may lead to a strong reduction of inconclusive pairs and provide some true hits without the need of actual geometry processing.

The use of *Web services* as seen in [18] was a powerful solution the authors found to reach scalability in an infrastructure designed to optimize query processing in distributed spatial databases. The scenario presented consists of independent organizations that produce data that may be overlapped geographically. This data is not intentionally replicated over the member nodes. A global index, based either on R-tree or Quadtree, is maintained in each node during all the time and if some localized data modification causes a change in its minimum bounding rectangle (MBR), all index replicas are updated. A query submitted to a node is then forwarded to other nodes that have data involved with it. The main goal was to reduce the traffic among nodes, improving queries response. The use of a global index distributed over the nodes had an important role in this context. The Web service approach, to easily turn the integration among nodes, and the involvement of several servers to deal with a query are relevant to our proposal.

The *Grid Greedy Node Scheduling Algorithm* (G2N) presented in [10] assigns sub-queries to grid nodes as a result of their throughputs expressed in tuples/second. If during runtime the actual throughput of a node differs too much from the previous known one, this new value is used to reschedule the remaining subqueries. A dynamic load balance, based on processors' throughput, is used in a similar manner on the SQB.

In [19] a dynamic load balancing strategy was adopted to address queries to grid clusters, adopting either a migration or a replication policy for the data, in order to explore the reliable nodes. The knowledge on the historical behavior of the clusters is a good heuristic to be explored when planning the query in the SQB.

OGSA-DQP [20], a Distributed Query Processor based on the OGSA architecture (a grid architecture based on services), has all the functionalities of a query broker: a *grid distributed query service* is responsible for compiling, optimizing, partitioning and scheduling distributed query execution plans over multiple execution nodes. Its *query evaluation service* is used by the previous service to execute a query plan. Some functional modules from the OGSA-DQP are being hosted by the SQB. The adoption of the OGSA-DAI, to permit uniform access in databases, is another common point between these architectures.

As pointed above, the spatial query broker's architecture proposed in this paper adopts some guidelines that were present in the previous works and is a consequence of our previous work [2,3].

6 Proposed Spatial Query Broker (SQB)

In this work a Spatial Query Broker is being proposed to cover the needs of a specialized mechanism to perform queries over SDBMS spread geographically and belonging to member organizations of a VO. The inability of traditional brokers to deal with databases as resources [16,15,21], particularly spatial databases, led us to research this topic. There was the need to incorporate in the SQB some functions that are specific of database management systems, like those related with the query processor. On the other hand, as a grid broker, it has to receive others skills from these specialized tools.

The main idea is to explore the skills from SDBMS members of the VO to perform some steps of spatial queries and, when necessary, divide the processing cost of the expensive ones with several grids' nodes, including those that are not specialized (without SDBMS), as depicted in figure 2.

The centralized approach seen in this figure: with central SQB, information service and replica location service are typical in some environments like the OGSA-DQP, Workload Management Service (WMS) and GridWay, these two last schedulers from gLite and Globus projects respectively. However, backup structures could coexist in virtual organizations in order to avoid a service break, in case of failure of the main module.

Following the gLite nomenclature conventions [22], each node capable of processing jobs is named Computing Element (CE). CEs in a virtual organization can have specific skills like high performance hardware or specialized services. The proposed architecture was tailored to treat spatial queries in a similar way as happens with common jobs: a query, after being typed by a user or passed by an application is guided to a specialized broker (spatial query broker).

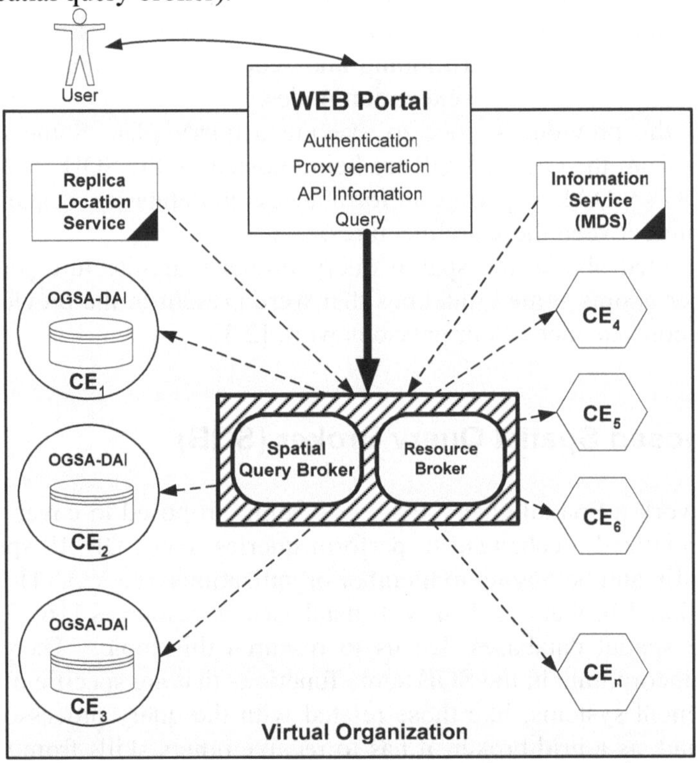

Fig. 2. VO composition with a Spatial Query Broker

The architecture (figure 3) can be presented based on a description of its component's modules as follows:

The *Coordinator* module has the role of manage all data flow since a query is received until it is finished. Additional tasks, such as checking a user's credential and starting a proxy session, are also performed by it.

The *Query Decomposition* module has almost all the functionalities found in traditional distributed query processing architectures [23]. The first of them is semantic analysis, based on a global schema, of the received query that can be performed by means of graph derived from the query and its analysis: when a node, representing a relation or a sub-graph is disconnected from the result, the query should be refused, since in this case some join predicate is missing. Another functionality is avoiding redundancy in the predicates, through simplifying both the geometric and non-geometric predicates, using the idempotent rules [23].

Locator module acts as a match-maker in a conventional broker [22]. Specifically, it is responsible for locate all sources of the themes mentioned in the received query, including their replicas and acquire statistical and metadata information like spatial data quality, databases' status, computing elements' status and their throughputs, besides of communication channels quality. The optimizer uses all this information in order to propose a good query plan and must be periodically updated in a grid information service (Globus MDS in this case). Replicas of databases are managed by the Replica Location Service (RLS), a service that maintain an index of all local replicas' catalogs, hosted on each CE, and permits queries about location of replicas related with a specific logical name.

An Optimizer has the task of dealing with the most common kinds of queries in a spatial database: window queries and spatial joins; based on this premise all its functioning aims at following a good strategy to reach the solution. In both cases the filter/refine strategy has proved to be a good one [6,5,7].

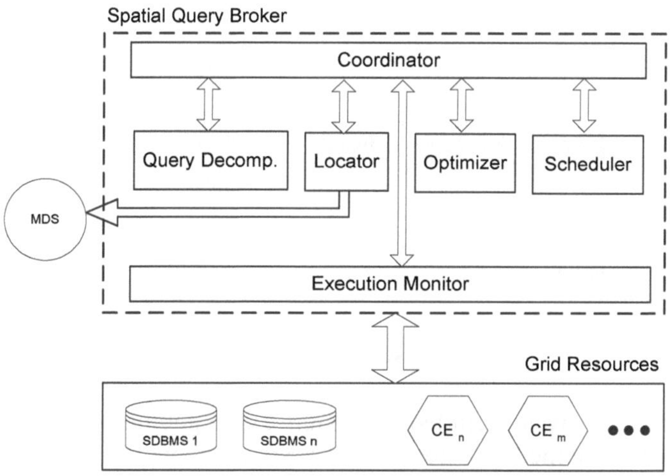

Fig. 3. Spatial Query Broker architecture

The *optimizer* takes the simplified query received and, based on the fragments' locations and status of replica-storing computing elements, chooses the best set of servers to execute the filtering phase of a spatial join operation. This phase is part of the multi-step filtering proposed in [6] and its main purpose is to discard false hits based on a MBR join operation followed by a geometric filtering like 4CRS [24]. The engine used to decide among the best servers makes use of the information collected from each of them in the *location module*. When dealing with either thematic or hybrid spatial data partitioning schema, the optimizer must request the transmission of the MBR approximations for the missing themes (and the 4CRS signature) to the specifics SDBMS, before proceeding with the filtering (figure 4). This procedure is well defined in [7].

The *Execution monitor* is responsible for submitting, to a set of servers defined in the previous module, the query plan received, and monitor its execution. After finishing their approximate sub-queries (MBR filtering and geometric filtering), the servers return two distinct sets of data to the execution monitor. The identifiers of the pairs of objects that attend to the intersection's predicate build the first set, and the pairs of inconclusive identifiers compose the second set with their vertices' numbers. It should be noticed that after executing the geometric filter phase based, for example, on 4CRS [25], some true hits may be already detected.

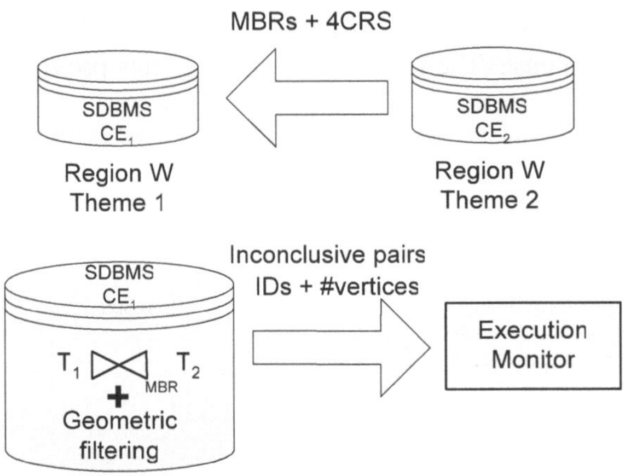

Fig. 4. Sub-queries' running sequence in CEs with SDBMS

A *scheduler module* was included to assign exact geometries' tests to common computing elements. These CEs receive inconclusive pairs to process as a result of their status, capability (throughput) and communication channels characteristics. The scheduler can adopt user's directives, collected from a web portal or from a file, to change the criteria used to sort the CEs, in order to receive the inconclusive pairs. The scheduler builds two queues with the inconclusive pairs, one with the pairs that have a total number of vertices above a threshold limit (supplied by the user) and the other with the other pairs. To proceed with the scheduling, each CE receives a pair of geometries to process (without alphanumeric data) as a result of its processing power (throughput) and new pairs are distributed as fast as they complete processing. The powerful CEs are fed with the pairs from the queue that stores those with greater number of vertices, while the other CEs receive their pairs from the other queue (figure 5).

Finishing an exact test, a CE informs the execution monitor whether that specific pair satisfies or not the intersection predicate. At the end, when all pairs have been processed, the execution monitor eliminates all redundant information and the coordinator orders the SDBMSs to transfer the tuples that satisfy the predicate to the requestor machine.

The dynamic behavior of the scheduling process permits the addressing of more complex geometries to more powerful CEs. This granularity based on pairs may lead to a fine adjustment when coupling geometries complexity with processor nodes.

A simplified sequence diagram of the entire SQB functioning is presented in figure 6, which can be described as follows:

- A user interacts with the *coordinator*, by means of a Web portal, submitting a query that has its themes chosen from a list (according to a global schema). This query should be constrained by a defined region;

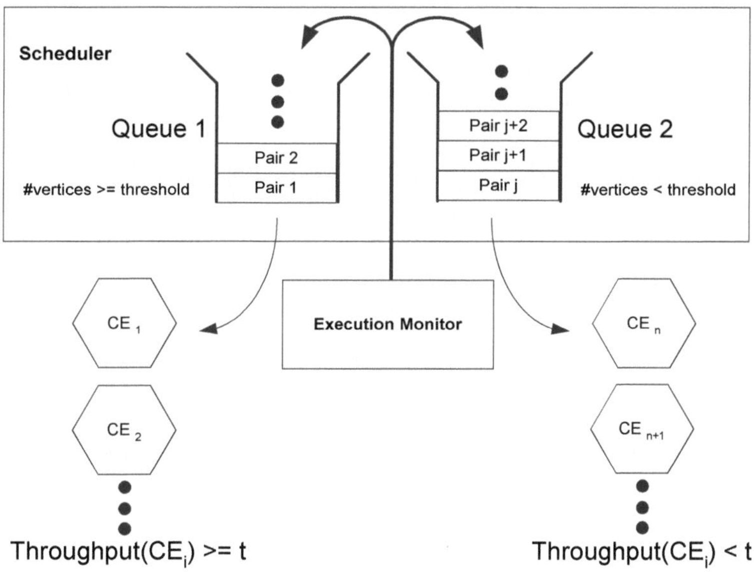

Fig. 5. Queues in the scheduler

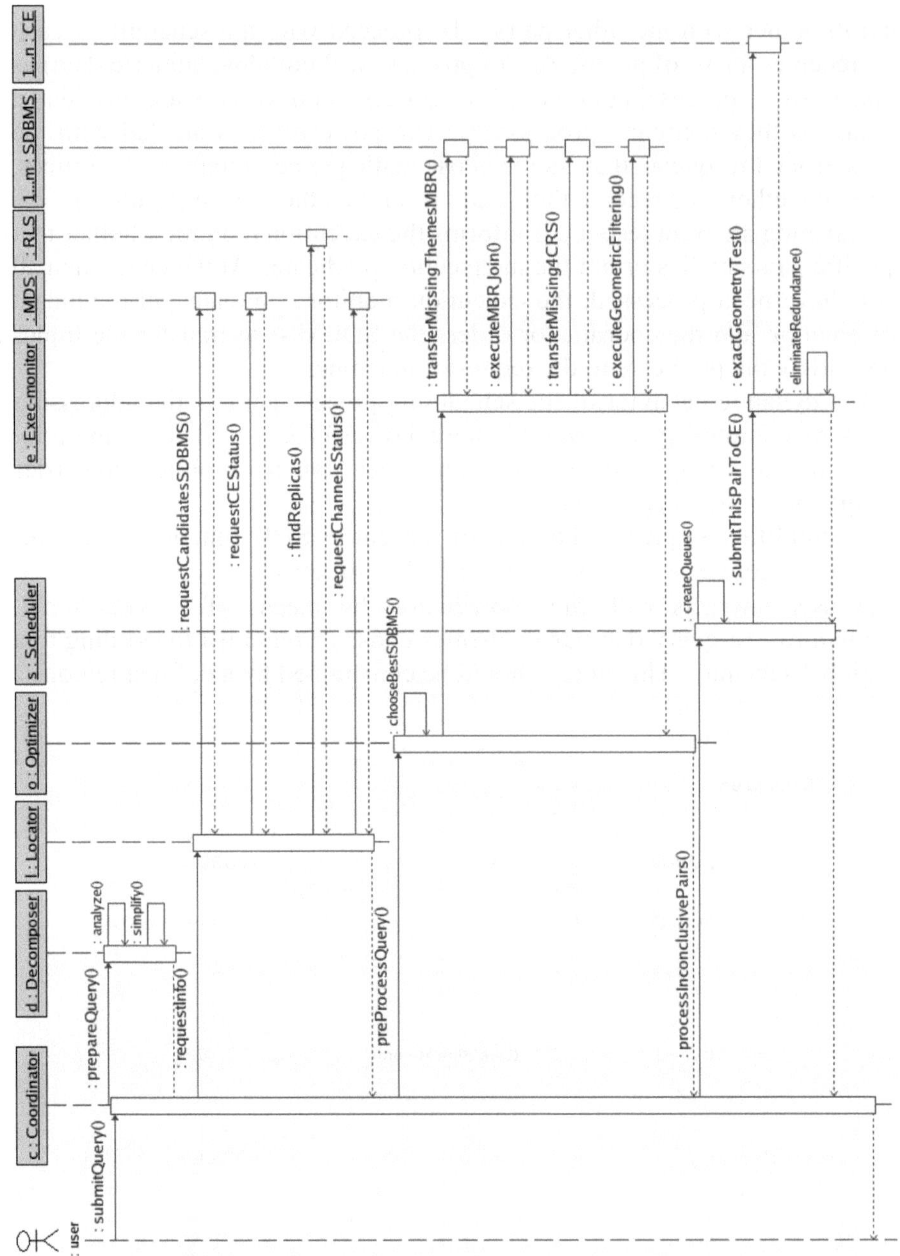

Fig. 6. Simplified sequence diagram of the SQB

- The *coordinator* calls a *decomposer* instance to analyze and simplify the query based on a global scheme and idempotent rules;
- If no problems occur the *coordinator* asks the *locator* to request the information service (MDS) for a list with the SDBMS that can participate of the query, based on the region covered by each one and the themes stored within them. Replicas of the databases associated with these themes are also requested to the RLS. The status of available unspecialized CEs, periodically updated by the MDS, and the network bandwidth between all these nodes and the SQB are also acquired;
- After all this information has been supplied the *coordinator* chooses the best SDBMSs to participate in the query and, by means of the *execution module*, determines the execution of the two initial steps of the Brinkhoff proposal [6]: the MBR join (filtering) and the geometric filtering. These two steps occur in a master CE, the one with the largest theme stored as suggested in [7];
- After receiving the results, formed by a list of pairs that satisfy the predicate and a list with inconclusive pairs, the *coordinator* generates the queues to store the sorted inconclusive pairs (only their ids and location) in order to send them to chosen CEs and proceed with the exact geometry tests;
- The results returned to the *execution module* are then combined.

It is desirable that all interactions between broker and SDBMS should take place under an OGSA-DAI interface, an implementation of data access and integration interface proposed by the Global Grid Forum, which allows dealing with a possible heterogeneity among these SDBMSs.

Complementing the description of the SQB components, it is necessary to depict some other premises considered in this proposal in order to clarify the reasons for the strategies chosen. The next paragraphs cover these aspects.

Queries being analyzed are restricted to spatial joins involving themes with polygon geometries under a set of common spatial predicates, like those defined by Egenhofer [26]: touch, overlap, inside and so on.

The family of spatial query joins covered by the present proposal can be seen in equation (*i*).

$$\text{Query} = \sigma_{W}(T_1) \underset{P\,=\,\text{intersection}}{\bowtie} \sigma_{W}(T_2) \qquad (i)$$

Where:

T_1 and T_2 are spatial themes,
W is a rectangular window and

P is the predicate used with the join operation.

There is also the premise that the VO offers some services needed by the SQB to interact with others VO's components. In the table 1 these services and their correspondent operations are listed.

Table 1. Other services needed in the VO

VO's Component	Service	Operations
Information Service	MDS	RequestGlobalSchema RequestCEStatus RequestCandidatesSDBMS RequestChannelsStatus RequestAGlobalID
CE (SDBMS)	OGSA-DAI	Many
	WS-GRAM	Many
	WSRF	ExactGeometryTest
CE	WS-GRAM	Many
	WSRF	ExactGeometryTest

The *ExactGeometryTest* service, deployed in CE nodes, has the task of receiving a pair of geometries with their IDs and execute the exact geometry processing to verify if they satisfy a predicate (intersection in this case). Its result is true or false according to the processing.

The last premise is that there is a global identification structure where each feature stored in the SDBMSs has been already registered. This mechanism makes possible for the components to deal with any feature no matter where it is stored. This structure is similar to a catalogue service.

This global ID mechanism is used during all broker activities and all new created features should be registered.

Most of the work mentioned in the previous section cannot be considered query brokers in a distributed environment: some just execute queries, without planning; others emphasize data migration and replication when scheduling queries, and so on. In table 2, the architecture proposed in this section is compared to some known brokers with the purpose of consolidating some aspects.

Table 2. Brokers comparison

	SQB	OGSA DQP	GridWay (Globus)	WMS (gLite)
Semantic	query	query	job	job
Job/query migration	no	no	Yes	no
Dynamic scheduling	yes	no	yes	no
Support to spatial queries	yes	no	-	-

Nodes without database involved with query	yes	no	-	-
Follow GGF standards	yes	yes	yes	yes

7 Experiments

A prototype is being built to validate the proposed architecture and many of the functionalities described are under construction. The Globus Toolkit is being adopted as the grid middleware since its basic tools are already robust and used in several projects around the world. Its job manager is responsible for receiving a job description and taking the necessary steps for its execution. This component is named Globus Resource and Allocation Manager (GRAM) and, in release 4, has the ability to deal with 32,000 concurrent jobs against 300 in the previous release [27].

The implementation is being made with a Web service approach by means of the Java WS-Core package provided by Globus. The SQB, accessed as a Web service in a specific server, makes use of stubs to interact with the others services like MDS, GRAM and the Replica Location Service (as depicted in figure 2) following a normal Web service style.

Secondo [28] was adopted as spatial database management system in this first stage. Its flexibility and modularity are achieved with an architecture based on algebras, which permit working with several data models. Spatial algebra is supplied with the product and changes or adjustments in its implementation can be easily done. The lack of an OGSA-DAI driver however, avoids its use in a heterogeneous group of servers. To test OGSA-DAI interface features, the PostgreSQL 8 / PostGIS will be adopted in a future stage.

Despite its being constructed, a few tests were done with synthetic spatial datasets consisting of polygons in order to give us some relative parameters to guide our work while dealing with spatial joins among polygons.

The tests were executed involving up to nine computers that run subqueries in parallel being the overall response time compared with that one obtained for a single machine running the entire query. With the times acquired, a speedup [29] parameter was obtained for each configuration.

The conditions adopted on tests are presented below:

- The spatial database servers were used to store regions of a regular grid, each of them with only two themes;
- The themes had their geometric attributes represented by triangles, that could vary in shape and size;

- Two datasets with 10,060 synthetic objects, each one, were partitioned in four and nine regular areas;
- Communication costs were estimated, since all tests were executed in a local area network in spite of using a remote network such as the Internet;
- Only nodes with SDBMS were involved in the query.

The queries assumed during the tests had the form:

Select all pairs
from theme1, theme2
where theme1 overlaps theme2 and region = regionX

The response time used to build the table 3 follows the equation (*ii*).

$$RT = T_{MSG} * \#messages + T_{TX} * \#bytes + T_{CPU} + T_{I/O} \qquad (ii)$$

It was considered during the tests that the number of messages and the time spent with a single message transmission were constant, so the first term from equation (*ii*) was not considered. The communication cost term in equation (*ii*) depends on size of data and on communication's channel bandwidth that was estimated in 256 kbps.

The final response time is limited by the node that computes the worst response time, following equation (*iii*).

$$RT_{FINAL} = \max\{RT_1, \ldots RT_i\} \; i = \text{num of servers working in parallel} \qquad (iii)$$

The results observed for the spatial joins are presented in table 3, where costs are expressed in milliseconds and the resultset size in bytes.

Table 3: CPU, I/O and communication costs

# of SDBMS	Region	# obj	Resultset size	Costs			Total Cost	Speed up
				CPU	I/O	Comm		
1 SDBMS	FULL	10912	2538685	2213416	735	77475	2291626	1.00
4 SDBMS	NW	2385	533445	133680	172	16279	150131	11.94
	NE	2706	628885	147175	194	19192	166561	
	SW	3154	734406	169332	233	22412	**191977**	
	SE	2747	640873	148509	209	19558	168276	
9 SDBMS	1	1065	246074	24794	65	7510	32369	45.07
	2	922	212832	24820	64	6495	31379	
	3	1167	269244	29273	85	8217	37575	
	4	1140	266527	28396	81	8134	36611	
	5	1338	312329	34590	97	9532	44219	
	6	1252	292853	29057	90	8937	38084	
	7	1440	336147	35398	96	10258	45752	
	8	1547	360394	39730	113	10998	**50841**	
	9	1138	265508	27332	81	8103	35516	

Figure 7 sketches the improvement achieved when executing the join operation in a parallel manner.

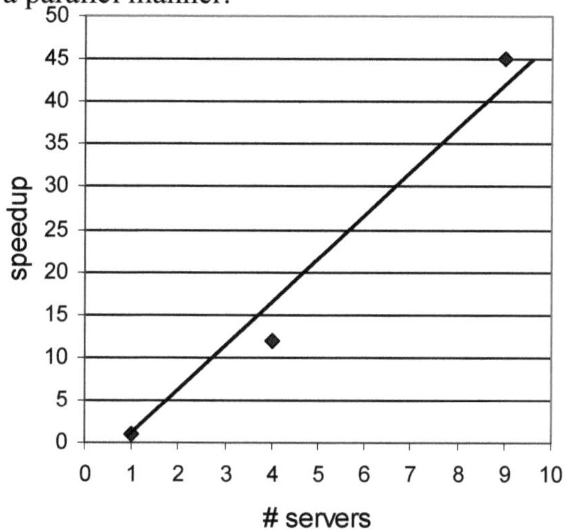

Fig. 7. Speedup

The results observed give us a small, but important contribution in the sense that we can perceive some aspects of the spatial fragmentation adopted with the proposed architecture.

When member organizations work like regional data producers, i.e., generating all themes on a spatial region, a global query can take advantage of pre-existing spatial indexes already created on each SDBMS and the original query can be easily broken into sub-queries. The executed tests fall in this category and show that:

- Processing (CPU plus I/O) cost of the spatial join queries is normally greater than communication costs when dealing with large window areas;
- Some objects are processed more than once because they cross the boundaries as presented in table 2 (the number of objects, when summed, is greater than the whole dataset);
- When subqueries are executed in parallel the ratio: *available memory / query complexity* increase, leading to a superlinear speedup. Superlinear speedup, as seen in table 2, means that the resources used to process the whole query at once in a server were insufficient and the operation consumed too much time.

We can conclude that expensive operations, like spatial joins, normally spend more time processing than transmitting data and, for this family of operations, we can notice expressive improvement in their response time depending of the number of machines involved in their execution, since the original query has the possibility to use an extra amount of free resources not available in a single machine.

However, an extra communication overhead can be expected when an excessive number of servers is involved with an operation. In these situations communication costs tend to be in the same order of magnitude or greater than processing costs, dominating overall response time. The effort made during the combination of partial results is another parameter that should be observed before a query subdivision, in order to avoid poorer performance.

Another point that should be emphasized in this structure is that the multi-processing of the same object must be avoided in order to improve the final response time. This problem causes great impact during the processing of exact geometries, since the algorithms' complexities, used to evaluate the queries' predicates, are directly dependent on their vertices' number. Besides that the amount of data to be transferred is increased, a post-processing stage to suppress redundant results being necessary.

8 Conclusion

The SQB architecture was conceived to explore the computational power of a virtual organization by means of dynamic scheduling in the most critical phase of a spatial join operation: the exact geometries processing. It is expected that queries over large spatial databases made up by several spatial database management systems, belonging to distinct organizations, can take advantage of this proposal, minimizing the effects of the natural grid's latency.

Government agencies responsible for the production of basic spatial data from large areas can benefit from it when offering query services over these huge amounts of data dynamically to other agencies, without the need of expensive clusters or equivalent equipment.

To summarize it, the adoption of a SQB in a grid environment avoids the need of a user knowing details of the locations of spatial data sources and gives him the possibility to run a query using the best resources available at that time. We suppose that the amount of computing elements running a query would have a strong impact on the final response time, as shown by the preliminaries tests. Another desirable feature of this architecture is that it explores the pre-existing spatial index of, at least, one of the two themes involved in the operation, reducing the time spent with the filtering and geometric filtering steps.

For the next stages we can highlight the prototype conclusion and the execution of several tests with actual and synthetic spatial datasets, with different parameter configurations such as vertex threshold, and with different cost models, the latter having a direct impact on the performance of the optimizer and of the scheduler.

References

[1] Câmara, G., Queiroz, G. (2002). GeoBR: Intercâmbio Sintático e Semântico de Dados Espaciais. In: IV Simpósio Brasileiro de Geoinformática,

[2] Meyer, W.S. et al Secondo-grid (2005) An Infrastructure to Study Spatial Databases in Computational Grids. In: VII Brazilian Symposium on Geoinformatics

[3] Meyer, W.S., Souza, J.M. (2006). Overlapped Regions with Distributed Spatial Databases in a Grid Environment. In: International Workshop on High-Performance Data Management in Grid Environments

[4] Smith, J. et al. (2002). Distributed Query Processing on the Grid.

[5] Hanssen, G. (2005). The Filter/Refine Strategy: A Study on the Land-Use Resource Dataset in Norway. In: SCANGIS

[6] Brinkhoff, T. et al (1994). Multi-Step Processing of Spatial Joins. In: ACM SIGMOD International Conference on Management of Data

[7] Ramirez, M.R. (2001). Spatial Distributed Query Processing. Paper presented at the COPPE/UFRJ, Rio de Janeiro, RJ

[8] Kang, M.-S., Choy, Y.-C. (2002). Deploying parallel spatial join algorithm for network environment. In: 5th IEEE International Conference on High Speed Networks and Multimedia Communications

[9] Foster, I. et al. (2001). The Anatomy of the Grid Enabling Scalable Virtual Organizations. Lecture Notes in Computer Science. 2150

[10] Porto, F. et al (2005). An adaptive distributed query processing grid service. In: Data Management in Grids

[11] Foster, I., Kesselman, C. (1999). The Grid: Blueprint for a New Computing Infrastructure. Computational grids, -Morgan-Kaufman

[12] Afgan, E. (2004). Role of the Resource Broker in the Grid. In: ACMSE

[13] Venegupal, S. et al (2004). A Grid Service Broker for Scheduling Distributed Data-Oriented Applications on Global Grids. In: 2nd Workshop on Middleware in Grid Computing, Toronto, Ontario, Canada

[14] Adzigogov, L. et al. (2005). EMPEROR: An OGSA Grid Meta-Scheduler based on Dynamic Resource. Journal of Grid Computing. 3, 19-37

[15] GridWay Team (2006). GridWay 5 Documentation: User Guide. Paper presented at the Universidad Complutense de Madrid, Madrid, Spain

[16] Andretto, P.e.a. (2004). Practical approaches to Grid workload and resource management in the EGEE project. In: EGEE-JRA1 Italian/Czech cluster meeting
[17] Di, L. et al (2003). The Integration of Grid Technology with OGC Web Services (OWS) in NWGISS for NASA EOS Data. In: GGF8 - The Eighth Global Grid Forum, Seattle, Washington, USA
[18] Ilya, Z. et al Online Querying of Heterogeneous Distributed Spatial Data on a Grid. In: The 3rd International Symposium on Digital Earth
[19] Mondal, A. et al. (2003). Effective Load-Balancing via Migration and Replication in Spatial Grids. Lecture Notes in Computer Science. 2736, 202-211
[20] UK Database Task Force OGSA-DQP 3.1 User's Documentation. Available in http://www.ogsadai.org.uk/documentation/ogsa-dqp_3.1/. Cited 10/7/2006
[21] Buyya, R., Venegupal, S. (2004). The Gridbus Toolkit for Service Oriented Grid and Utility Computing: An overview and Status Report. In: First IEEE Intl Workshop on Grid Economics and Business Models, Seoul, Korea
[22] EGEE (2006). GLite - Installation and Configuration Guide v 3.0 (rev 2). Paper presented at the European Union
[23] Özsu, M.T., Valduriez, P. (2001). Principles of Distributed Database Systems. Prentice-Hall
[24] Azevedo, L.G. et al (2004). Approximate Spatial Query Processing Using Raster Signature. In: Brazilian Symposium on Geoinformatics
[25] Zimbrão, G., Souza, J.M. (1998). A Raster Approximation for the Processing of Spatial Joins. In: 24th VLDB
[26] Egenhofer, M.J., Herring, J.R. (1994). Categorizing Binary Topological Relations Between Regions, Lines and Point in Geographical Databases. Paper presented at the NCGIA
[27] Foster, Ian Globus Toolkit 4. Available in www.gridbus.org/escience/051205GlobusTutorialeScience.ppt. Cited 22/7/2006
[28] Güting, R.H. et al. (2004). Secondo: An Extensible DBMS Architecture and Prototype. Paper presented at the Fernuniversität Hagen, Hagen, Germany
[29] Gistafson, J.L. (1990). Fixed Time, Tiered Memory, and Superlinear Speedup. In: Fifth Distributed Memory Computing Conference (DMCC5)

TerraHS: Integration of Functional Programming and Spatial Databases for GIS Application Development

Sérgio Souza Costa, Gilberto Câmara, Danilo Palomo

Divisão de Processamento de Imagens
Instituto Nacional de Pesquisas Espaciais

1 Introduction

Recent, research in GIScience proposes to use functional programming for geospatial application development [1-5]. Their main argument is that many of theoretical problems in GIScience can be expressed as algebraic theories. For these problems, functional languages enable fast development of rigorous and testable solutions [2]. However, developing a GIS in a functional language is not feasible, since many parts needed for a GIS are already avaliable in imperative languages such as C++ and Java. This is especially true for spatial databases, where applications such as PostGIS/PostgreSQL offer a basic support for spatial data management. It is unrealistic to develop such support using functional programming.

It is easier to benefit from functional programming for GIS application development if we build an application on top of an existing spatial database programming environment. This work presents TerraHS, an application that enables developing geographical applications in a functional language, using the data handling provided by TerraLib. TerraLib is a C++ library that supports different spatial database management systems, and that includes many spatial algorithms. As a result, we get a combination of the good features of both programming paradigms.

This paper describes the TerraHS application. We briefly review the literature on functional programming and its use for GIS application development in Section 2. We describe how we built TerraHS in Section 3. In Section 4, we show the use of TerraHS for developing a Map Algebra.

2 Brief Review of the Literature

2.1 Functional Programming

Functional programming is a programming paradigm that considers that computing is evaluating off mathematical functions. Functional programming stresses functions, in contrast to imperative programming, which stresses changes in state and sequential commands [6]. Recent functional languages include Scheme, ML, Miranda and Haskell. TerraHS uses the Haskell programming language. The Haskell report describes the language as:

> *"Haskell is a purely functional programming language incorporating many recent innovations in programming language design. Haskell provides higher-order functions, non-strict semantics, static polymorphic typing, user-defined algebraic datatypes, pattern-matching, list comprehensions, a module system, a monadic I/O system, and a rich set of primitive datatypes, including lists, arrays, arbitrary and fixed precision integers, and floating-point numbers"* [7].

The next section provides a brief description of the Haskell syntax. This description will help the reader to understand the essential arguments of this paper. For detailed description of Haskell see [7], [8] and [9].

2.2 A Brief Tour of the Haskell Syntax

Functions are the core of Haskell. A simple example is a function that which adds its two arguments:

```
add :: Integer → Integer → Integer
add x y =  x + y
```

The first line defines the `add` function. It takes two `Integer` values as input and produces a third one. Functions in Haskell can also have generic (or polymorphic) types. For example, the following function calculates the length of a generic list, where `[a]` is a list of elements of a generic type `a`, `[]` is the empty list, and `(x:xs)` is the list composition operation:

```
length :: [a] → Integer
length [] = 0
length (x:xs) = 1 + length xs
```

This definition reads "`length` is a function that calculates an integer value from a list of a generic type `a`. Its definition is recursive. The length of an empty list is zero. The length of a nonempty list is one plus the length of the list without its first element".

The user can define new types in Haskell using a `data` declaration, which defines a new type, or the `type` declaration, which redefines an existing type. For example, take the following definitions:

```
type Coord2D = (Double, Double)
data Point       = Point Coord2D
data Line2D      = Line2D [Coord2D]
```

In these definitions, a `Coord2D` type is shorthand for a pair of `Double` values. A Point is a new type that contains one `Coord2D`. A `Line2D` is a new type that contains a list of `Coord2D`. One important feature of Haskell lists is that they can be defined by a mathematical expression similar to a set notation. For example, take the expression:

```
[elem | elem <- (domain map) , (predicate elem obj)]
```

It reads "the list contains the elements of a map that satisfy a predicate that compares each element to a reference object". This expression could be used to select all objects that satisfy a topological operator ("*all roads that cross a city*"). Haskell includes *higher-order* functions. These are functions that have other functions as arguments. For example, the `map` higher-order function applies a function to a list, as follows:

```
map    :: (a→b) → [a] → [b]
map f  []       =  []
map f (x:xs)    =  f x : map f xs
```

This definition can reads as "take a function of type `a→b` and apply it recursively to a list of `a`, getting a list of `b`". Haskell supports overloading using *type classes*. A definition of a *type class* uses the keyword `class`. For example, the type class `Eq` provides a generic definition of all types that have an equality operator:

```
class Eq a where
   (==)   :: a → a → Bool
```

This declaration reads "a type `a` is an instance of the class `Eq` if it defines is an overloaded equality `(==)` function." We can then specify instances of type class `Eq` using the keyword `instance`. For example:

```
instance Eq Coord2D where
  ((x1,x2) == (y1,y2)) = (x1 == x2 && y1 == y2)
```

Haskell also supports a notion of *class extension.* For example, we may wish to define a class `Ord` which *inherits* all the operations in `Eq`, but in addition includes comparison, minimum and maximum functions:

```
class (Eq a) => Ord a  where
     (<), (<=), (>=), (>)   :: a → a → Bool
     max, min               :: a → a → a
```

2.3 Functional Programming and GIS

Many recent papers propose using functional languages for GIS application develepment [1-3, 5]. Frank and Kuhn [2] show the use of functional programming languages as tools for specification and prototyping of Open GIS specifications. Winter and Nittel [5] apply a formal tool to writing specifications for the Open GIS proposal for coverages. Medak [4] develops an ontology for life and evolution of spatial objects in an urban cadastre. To these authors, functional programming languages satisfy the key requirements for specification languages, having expressive semantics and allowing rapid prototyping. Translating formal semantics is direct, and the resulting algebraic structure is extendible. However, these works do not deal with issues related to I/O and to database management. Thus, they do not provide solutions applicable to real-life problems. To apply these ideas in practice, we need to integrate functional and imperative programming.

2.4 Integration of Functional and Imperative Languages

The integration functional and imperative languages is discussed in Chakravarty [10], who presents the *Haskell 98 Foreign Function Interface (FFI),* which supports calling functions written in C from Haskell and vice versa. However, functions written in imperative languages can contain side effects. To allow functional languages to deal with side effects, Wadler [11] proposed *monads* for structuring programs written in functional language. The use of monads enables a functional language to simulate an imperative behavior with state control and side effects [9]. Jones [12] presents many crucial issues about interaction of functional languages with the external world, such as I/O, concurrency, exceptions and interfaces to libraries written in other languages. In this work, the author describes a Haskell web server as a case study. These works show of the integration between these two programming styles. However, none of these works deals with geoinformation systems. On the next section we present an application that integrates programs written in Haskell with spatial databases and allows fast and reliable GIS application development.

3 TerraHS

This section presents TerraHS, a software application which enables developing geographical applications using in functional programming using data stored in a spatial database. TerraHS links the Haskell language to in the TerraLib GIS library. TerraLib is a class library written in C++, whose functions provide spatial database management and spatial algorithms. TerraLib is free software [13]. TerraHS links to the TerraLib functions through the Foreign Function Interface [10] and a function set written in C language, which performs the TerraLib functions. The Figure 1 shows its architecture.

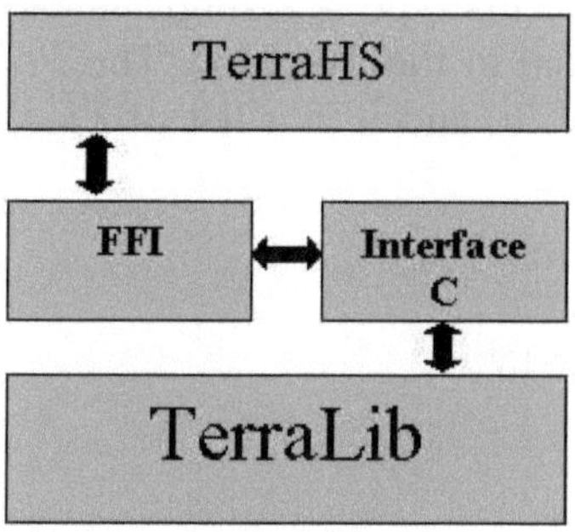

Fig. 1. TerraHS Architecture

TerraHS includes three basic resources for geographical applications: spatial representations, spatial operations and database access. The next sections present them.

3.1 Spatial Representations

3.1.1 Vector data structures

Identifiable entities on the geographical space, or *geo-objects*, such as cities, highways or states are usually represented in vector data structures, as point, line and polygon. These data structures represent an object by one or more pairs of Cartesian coordinates. TerraLib represents coordinate pairs through the *Coord2D* data type. In TerraHS, this type is a tuple of real values.

```
type Coord2D = (Double, Double)
```

The type Coord2D is the basis for all the geometric types in TerraHS, namely:

```
data Point      = Point Coord2D
data Line2D     = Line2D [Coord2D]
type LinearRing = Line2D
data Polygon    = Polygon [LinearRing]
```

The *Point* data type represents a point in TerraHS, and is a single instance of a *Coord2D*. The *Line2D* data type represents a line, composed of one or more segments and it is a vector of *Coord2Ds* [13]. The *LinearRing* data type represents a closed polygonal line. This type is a single instance of a *Line2D*, where the last coordinate is equal to the first [13]. The *Polygon* data type represents a polygon in TerraLib, and it is a list of *LinearRing*. Other data types include:

```
data PointSet   = PointSet [Point]
data LineSet    = LineSet [Line2D]
data PolygonSet = PolygonSet [Polygon]
```

3.1.2 Cell-Spaces

TerraLib supports cell spaces. Cell spaces are a generalized raster structure where each cell stores a more than one attribute value or as a set of polygons that do not intercept one another. A cell space enables joint storage of the entire set of information needed to describe a complex spatial phenomenon. This brings benefits to visualization, algorithms and user interface [13]. A cell contains a bounding box and a position given by a pair of integer numbers.

```
data Cell = Cell Box Integer Integer
data Box  = Box Double Double Double Double
```

The Box data type represents a bounding box and the Cell data type represents one cell in the cellular space. The CellSet data type represents a cell space.

```
data CellSet = CellSet [Cell]
```

3.1.3 Spatial Operations

TerraLib provides a set of spatial operations over geographic data. TerraHS provides function that use those algorithms. We used Haskell type classes [14, 15] to define the spatial operations using polymorphism. These topologic operations can be applied for any combination of types, such as point, line and polygon.

```
class TopologyOps a b where
   disjoint :: a → b → Bool
   intersects :: a → b → Bool
   touches  ::  a → b → Bool
   ...
```

The TopologyOps class defines a set of generic operations, which can be instantiated to several combinations of types:

```
instance TopologyOps Polygon Polygon
instance TopologyOps Point Polygon
instance TopologyOps Point Line2D
...
```

3.2 Database Access

One of the main features of TerraLib is its use of different object-relational database management systems (OR-DBMS) to store and retrieve the geometric and descriptive parts of spatial data [13]. TerraLib follows a layered model of architecture, where it plays the role of the middleware between the database and the final application. Integrating Haskell with TerraLib enables an application developed in Haskell to share the same data with applications written in C++ that use TerraLib, as shown in Figure 2.

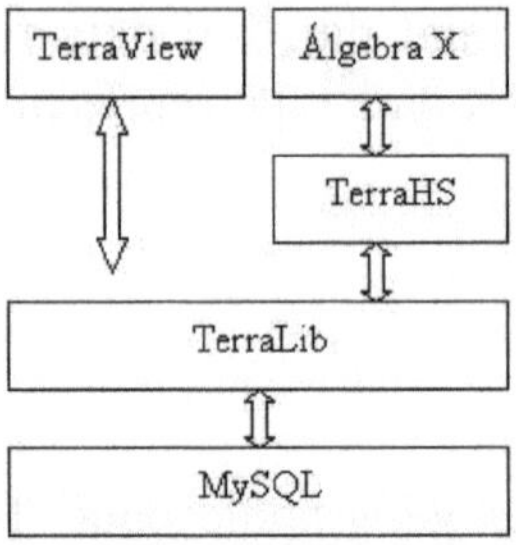

Fig.2. Using the TerraLib to share a geographical database.

A TerraLib database access does not depends on a specific DBMS and uses an abstract class called *TeDatabase* [13]. In TerraHS, the database classes are algebraic data types, where each constructor represents a sub-class.

```
data Database = MySQL String String String String
  | PostgreSQL String String String String
```

A TerraLib *layer* aggregates spatial information located over a geographical region and that share the same attributes. A layer is identifier in a TerraLib database by its name [13].

```
type LayerName = String
```

In TerraLib, a *geo-object* is an individual entity that has geometric and descriptive parts, composed by:

- **Identifier**: identifies a *geo-object*.

```
data ObjectId = ObjectId String
```

- **Attributes**: this is the descriptive part of a geo-object. An attribute has a name (*AttrName*) and a value (*Value*).

```
type AttrName = String
data Value = StValue String| DbValue Double
           |InValue Int | Undefined
data Atribute = Atr (AttrName, Value)
```

- **Geometries**: this is the spatial part, which can have different representations.

```
data Geometry = GPt Point | GLn Line2D
      | GPg Polygon |GCl Cell (…)
```

A *geo-object* in TerraHS is a triple:

```
data GeObject=GeoObject(ObjectId,[Atribute],Geometry)
```

The *GeoDatabases* type class provides generic functions for storage, retrieval of geo-objects from a spatial database.

```
class GeoDatabases a where
  open :: a → IO (Ptr a)
  close :: (Ptr a) → IO ()
  retrieve :: (Ptr a) → LayerName → IO [GeoObject]
  store::(Ptr a)→LayerName→[GeoObject]→ IO Bool
  errorMessage :: (Ptr a) → IO String
```

These operations will then be instantiated to a specific database, such as `mySQL` or `PostgreSQL`. Figure 3 shows an example of a TerraLib database access program.

```
host = "sputnik"
user = "Sergio"
password = "terrahs"
dbname = "Amazonia"
main:: IO()
main = do
    -- accessing TerraLib database
   db <- open (MySQL host user password dbname)
    -- retrieving a geo-object set
   geos <- retrieve db "cells"
   geos2 <- op geos -
    -- storing a geo-object set
   store db "newlayer" geos2
   close db
```

Fig.3. Acessing a TerraLib database using TerraHS

4 A generalized map algebra

One of the important uses of functional language for GIS is to enable fast and sound development of new applications. As an example, this section presents a map algebra in a functional language. In GIS, *maps* are a continuous variable or to a categorical classification of space (for example, soil maps). Map Algebra is a set of procedures for handling maps. They allow the user to model different problems and to get new information from the existing data set. The main contribution to map algebra comes from the work of Tomlin [16]. Tomlin's model uses a single data type (a map), and defines three types of functions. *Local* functions involve matching locations in different map layers, as in "*classify as high risk all areas without vegetation with slope greater than 15%*". *Focal* functions involve proximal locations in the same layer, as in the expression "*calculate the local mean of the map values*". *Zonal* functions summarize values at locations in a layer contained in zones defined in another layer. An example is "*given a map of city and a digital terrain model, calculate the mean altitude for each city.*"

For this experiment, we use the map algebra proposed in Câmara et al. [17]. The authors describe the design of a map algebra that generalizes Tomlin's map algebra by incorporating topological and directional spatial predicates. In the next section, we describe and implement this algebra.

4.2 The map abstract data type

Our map algebra has two main data types: object set and field. An *object set* is a set of objects represented by points, lines or regions associated with nonspatial attribute. *Fields* are functions that map a location in a spatial partition to a nonspatial attribute. The *map* data type combines both the *object set* data type and the *field* data type. A map is a function $m{:}{:}\ E \rightarrow A$, where:

- The domain is finite collection, either a set of cells or a set of objects.
- The range is a set of *attribute values*.

For each geographic element $e \in E$, a *map* returns a value $m\ (e) = a$, where $a \in A$. A geographical element can represent a location, area, line or point. This definition matches the definition of a *coverage* in Open GIS [18]. A coverage in a planar-enforced spatial representation that *covers* a geographical area completely and divides it in spatial partitions that may be either regular or irregular. For retrieving data from a coverage, the Open GIS specification propose describes a discrete function (*DiscreteC_Function),* as shown in Figure 4 below.

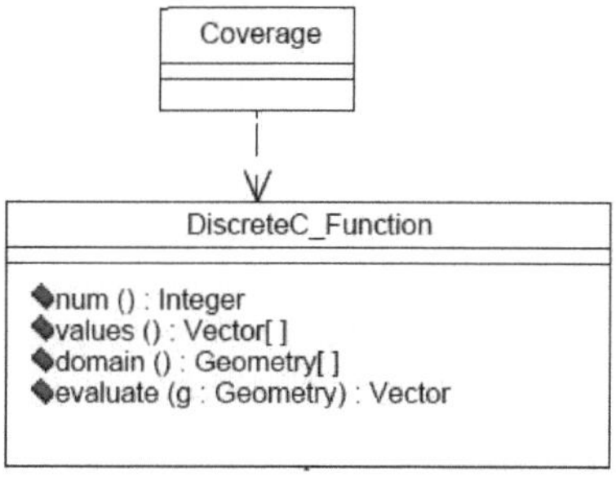

Fig.4. The Open GIS discrete coverage function – source: [18].

The *DiscreteCFunction* data type describes a function whose spatial domain and whose range are finite. The domain consists of a finite collection of geometries, where a *DiscreteCFunction* maps each geometry for a value [18]. Based on the Open GIS specification, we defined the type class *Maps*. The type class *Maps* generalizes and extends the *DiscreteCFunction* class. Its functions are parameterized on the input type `a` and the output type `b`. It provides the support for the operations proposed by the *DiscreteCFunction*:

```
class Maps map where
  evaluate :: (Eq a, Eq b) => map a b → a → Maybe b
  domain   :: map a b → [a]
  num      :: map a b → Int
  values   :: map a b → [b]
  new_map  :: [a] → (a → b ) → (map a b)
  fun      :: (map a b) → (a → b)
```

The functions is the Maps type class work as follows: (a) `evaluate` is a function that takes a `map` and an input value `a` and produces an output value ("*give me the value of the map at location a*"); (b) `domain` is a function that takes a map and returns the values of its domain; (c) `num` returns the number of elements of the map's domain; (d) `values` returns the values of the map's range. We propose two extra functions: *new_map* and *fun*, as described below.

- `new_map`, a function that returns a new `map` *m,* given a domain and *a coverage function.*
- `fun`: given a `map`, returns its *coverage function.*

We defined the *Map* data type to use the functions of the generic type class *Maps*. The *Map* data type is also parameterized.

```
data Map a b = Map ((a → b), [a])
```

The data type *Map* has two parts:

- A *coverage function* that maps an object of generic type *a* to generic type *b*.
- A domain of objects of the polymorphic type *a*.

The instance of the type class *Maps* to the *Map* data type is shown below:

```
instance Maps Map where
  new_map a f = (Map (f, a))
  evaluate f o
       |(elem o (domain f)) = Just ((fun f) o)
       |otherwise = Nothing
  domain (Map (f, a)) = a
  num f = length (domain f)
  values f  = map (fun f) (domain f)

    fun (Map (f,_)) = f
```

Figure 5 show an example of the Map data type.

```
m1 :: (Map String Integer)
m1 = new_map ["ab","abc","a"] length
values m1
= [2,3,1]
evaluate m1 "ab"
= Just 2
evaluate m1 "ad"  -- m1 not contain "ad"
= Nothing
```

Fig.5. Example of use of the *Map* data type.

4.2 Operations

Câmara et al [19] define two classes of the map algebra operations: nonspatial and spatial. For *nonspatial operations*, the value of a location in the output map is obtained from the values of the same location in one or more input maps. They include logical expressions such as *"classify as high risk all areas without vegetation with slope greater than 15%"*, *"Select areas higher than 500 meters"*, *"Find the average of deforestation in the last two years"*, and *"Select areas higher than 500 meters with temperatures lower than 10 degrees"*. *Spatial functions* are those where the value of a location in the output map is computed from the values of the neighborhood of the same location in the input map. They include expressions such as *"calculate the local mean of the map values"* and *"given a map of cities and a digital terrain model, calculate the mean altitude for each city"*. In what follows, we show these operations in TerraHS, using polymorphic data types.

4.2.3 Nonspatial operations

Nonspatial operations are higher-order functions that take one value for each input map and produce one value in the output map, using a first-order function as argument. These include *single argument functions* and *multiple argument functions* [19].

```
class (Maps m) => NonSpatialOperations m where
  map_single   :: (b → c) → (m a b) → (m a c)
  map_multiple::([b]→ c)→[(m a b)]→(m a b)→(m a c)
```

The *map_single* function has two arguments: a *map m* and a first-order function *g*. It returns a new *map*, whose domain contains the same elements of the input map domain. The *coverage function* of the output map is the composition of the *coverage function* of the input map *m* and the first-order function *g*.

```
map_single g m = new_map (domain m)   ( g . (fun m))
```

Defines a new map with the same domain	defines the mapping function of the new map

Figure 6 shows an example of a single argument function.

```
values m1
= [2, 4, 12]
m2 = map_single square m1
values m2
= [4, 16, 144]
```

Fig.6. Example of use of the single argument function

The *map_multiple* function has three arguments: a *map* list, a multivalued function and a reference *map*. Given a reference *map,* it applies a multivalued function in *map* list.

```
map_multiple fn mlist mref =
new_map (domain mref) (\x → fn (map_r mlist x))
```

defines a new map with the same domain		defines the mapping function of the new map using an auxiliary function

The *map_multiple* function returns a new *map* with a same domain of the reference map and a new coverage function. This function uses the auxiliary function *map_r*. For each element *x* of the reference map, *map_r* applies the multiargument function in the input list of maps to get the output value. It also handles cases where there are multiargument function fails to returns an output value.

```
map_r :: (Maps m) => [(m a b )] → a → [b]
  map_r [] _ = []
  map_r (m:ms) e = map_r' (evaluate m x)
  where
      map_r' (Just v)   = v : (map_r ms e)
      map_r' (Nothing) = (map_r ms e)
```

Figure 7 shows an example of *map_multiple*. In this example, the *m3 map* is the result of the sum of the *maps m1* and *m2*.

```
values m1
= [2, 4, 8]
values m2
= [4, 5, 10]
m3 = map_multiple sum [m1, m2] m1
values m3
= [6, 9, 18]
```

Fig.7. Example of use of *map_multiple*

4.2.2 Spatial Operations

Spatial operations are higher-order functions that use a spatial predicate. These functions combine a selection function and a multivalued function, with two input maps (the reference map and the value map) and an output map [19]. Spatial functions generalize Tomlin's focal and zonal operations and have two parts: *selection* and *composition*. For each location in the output map, the *selection function* finds the matching region on the reference map. Then it applies the spatial predicate between the *reference* map and the *value* map and creates a set of values. The *composition function* uses the selected values to produce the result (Figure 8). Take the expression *"given a map of cities and a digital terrain model, calculate the mean altitude for each city".* In this expression, the *value map* is the digital terrain model and the *reference map* is the map of cities. The evaluation has two parts. First, it selects the terrain values inside each city. Then, it calculates the average of these values.

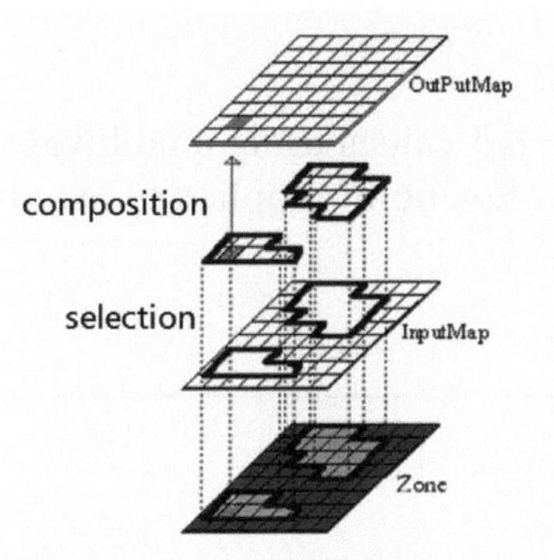

Fig.8. Spatial operations (selection + composition). Adapted from [16].

The implicit assumption is that the geographical area of the *output map* is the same as *reference map*. The type signature of the spatial functions in TerraHS is:

```
class (Maps m) => SpatialOperations m where
  map_select::(m a b)→(a → c→ Bool)→ c → (m a b)
  map_compose :: ([b] → b) → (m a b) →  b
  map_spatial::([b]→b) → (m a b) → (a → c → Bool)
    → (m c b) → (m c b)
```

The *spatial selection* function selects all elements that satisfy a predicate on a reference object ("*select all deforested areas inside the state of Amazonas*"). It has three arguments: an input *map*, a predicate and a reference element.

```
map_select m pred obj  = new_map sel_dom (fun m)
  where
  sel_dom=[elem|elem ← (domain m) , (pred elem obj)]
```

This function takes a reference element and an input map. It creates a map that contains all elements of the input map that satisfy the predicate over the reference element. Figure 9 shows an example, where the map consists of a set of points. Then, we select those points that intersect a given line.

```
ln =Line2D[Point(1,2),Point(2,2),Point(1,3),Point(0,4)]
domain m1
= [Point(4,5),Point (1,2),Point (2,3),Point (1,3)]
m2 = map_select m1 intersects ln
domain m2
= [Point (1,2), Point (1,3)]
```

Fig.9. Example of *map_select.*

The *composition function* combines selected values using a multivalued function. In Figure 10, the `map_compose` function is applied to *map m1* and to the multivalued function *sum*.

```
map_compose f m = (f (values m))
```

```
values m1
= [ 2, 6, 8]
map_compose sum m1
= 16
```

Fig.10. Example of *map_compose.*

The `map_spatial` function combines *spatial selection* and *spatial composition*:

```
map_spatial fn m pred mref = new_map (domain mref)
  (\x → map_compose (map_select m pred x) fn)
```

Map_spatial creates a map whose domain contains the elements of the reference map. To get its coverage function, we apply *map_compose* to the result of the *map_selection.* Figure 11 shows an example.

```
domain m1
=[Point(4,5),Point (1,2),Point (2,3),Point (1,3)]
values m1
=[2,4,5,10]
domain m2
=[(Line2d[Point(1,2),Point(2,2),Point(1,3),Point(0,4)])]
m3=map_spatial sum m1 intersects m2
values m3     -- 4 + 10
= [14]
```

Fig. 11. Example of map_spatial

The spatial operation selects all points of *m1* that intersect *m2* (which is a single line). Then, it sums its values. In this case, points (1,2) and (1,3) intersect the line. The sum of their values is 14.

4.3 Application Examples

In the previous section we described how to express the map algebra proposed in Câmara et al. [19] in TerraHS. In this section we show the application of this algebra to actual geographical data.

4.3.1 Storage and Retrieval

Since a *Map* is generic data type, it can be applied to different concrete types. In this section we apply it to the *Geometry* and *Value* data types available in the TerraHS, which represent, respectively, a region and a descriptive value. TerraHS enables storage and retrieval of a *geo-object* set. To perform a map algebra, we need to convert from a *geo-object* set to a map and vice versa.

```
toMap::[GeObject]→ AttrName→ (Map Geometry Value)
toGeObject::(Map Geometry Value)→AttrName→[GeObject]
```

Given a geo-object set and the name of one its attributes, the `toMap` function returns a `map`. Remember that a *Map* type has one value for each region. Thus, a layer with three attributes it produce three *Maps*. The `toGeObject` function inverts the `toMap` function. Details of these two functions are outside the scope of this paper. Given these functions, we can store and retrieve a map, given a spatial database.

```
retrieveMap::
    Database → LayerAttr → IO (Map Geometry Value)
  retrieveMap db (layername, attrname) = do
      db <- open db
      geoset <- retrieve db layername
      let      map = toMap geoset attrname
      close db
      return map
```

The `LayerAttr` type is a tuple that represents the layer name and attribute name. The `retrieveMap` function connects to the database, loads a geo-object set, converts these geo-objects into a map, and return this map as its output.

```
storeMap:: Database →
   LayerAttr→(Map Geometry Value)→IO Bool
storeMap db (layername, attrname) m = do
   let geos = toGeObject map attrname
   db <- open db
   close db
   let status = store db layername geos
   return status
```

The `storeMap` function coverts a map to a geo-object set that will be saved in the database. We can now write a program that reads and writes a *map* in a TerraLib database.

```
host = "sputnik"
user = "Sergio"
password = "terrahs"
dbname = "Amazon"
main:: IO ()
main = do
    db <- open (MySQL host user pass dbname)
   def_map<-retrieveMap db ("amazonia","deforest")
   -- apply a nonspatial operation
    let defclass = map_single classify def_map
    storeMap db ("amazon", "defclass")  defclass
```

Fig.12. Retrieving and storing a *Map* from TerraLib Database

4.3.2 Examples of Map Algebra in TerraHS

Since 1989, the Brazilian National Institute for Space Research has been monitoring the deforestation of the Brazilian Amazon, using remote sensing images. We use some of this data as a basis for our examples. We selected a data set from the central area of Pará, composed by a group of highways and two protection areas. This area is divided in cells of 25 x 25 km2, where each cell describes the percentage of deforestation and deforested area (Figure 13).

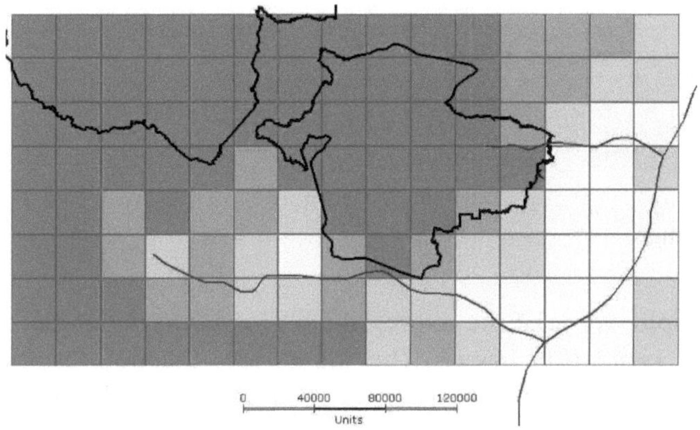

Fig.13. Deforestation, Protection Areas and Roads Maps (Pará State)

Our first example considers the expression: "*Given a map of deforestation and classification function, return the classified map*". The classification function defines four classes: (1) dense forest; (2) mixed forest with agriculture; (3) agriculture with forest fragments; (4) agricultural area. This function is:

```
classify :: Value → Value
classify (DbValue v)
     | v < 0.2 = (StValue "1")
     | ((v > 0.2) && (v < 0.5)) = (StValue "2")
     | (v > 0.5) && (v < 0.8) = (StValue "3")
     | v > 0.8 = (StValue "4")
```

We obtain the classified map using the `map_single` operation together with the `classify` function:

```
..def_class = map_single classify def_map
```

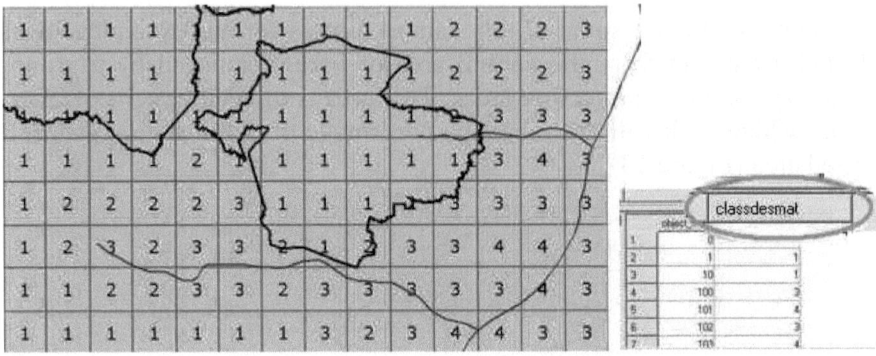

Fig.14. The classified map

As a second example, we take the expression: *"Calculate the mean deforestation for each protection area"*. The inputs are: the deforestation map (`def_map`), a spatial predicate (`within`), a multivalued function (`mean`) and the map of protected areas (`prot_areas`). The output is a deforestation map of the protected areas (`def_prot`) with the same objects as the reference map (`prot_areas`). We use the `map_spatial` higher-order operation to produce the output:

```
def_prot=map_spatial mean def_map within prot_areas
```

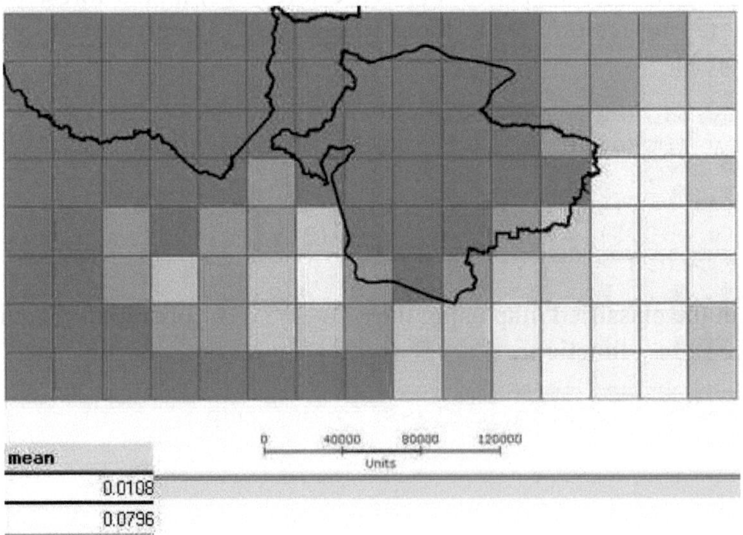

Fig.15. Deforest mean by protection area

In our third example, we consider the expression: "*Given a map containing roads and a deforestation map, calculate the mean of the deforestation along the roads*". We have as inputs: the deforestation map (`def_map`), a spatial predicate (`intersect`), a multivalued function (`mean`) and a road map (`roads`). The product is a map with one value for each road. This value is the mean of the cells that intercept this road.

```
road_def=map_spatial mean def_map intersect road_map
```

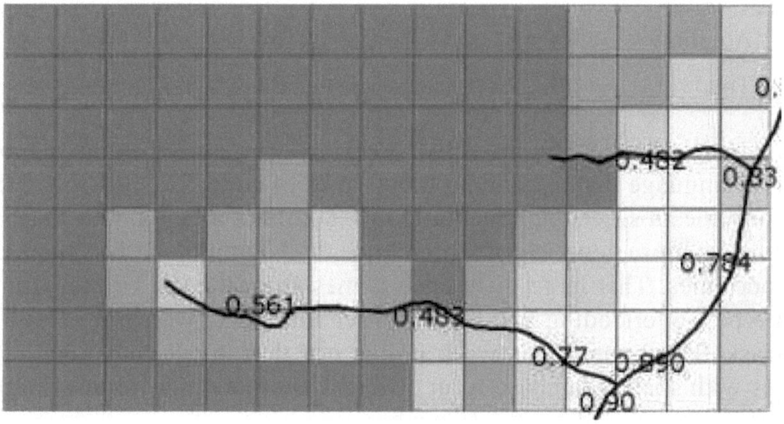

Fig. 16 Deforestation mean along the roads

5 Conclusions

This paper presents the TerraHS application for integrating functional programming and spatial databases. We use TerraHS to develop and validate a map algebra in a functional language. The resulting map algebra is compact, generic and extensible. The example shows the benefits on using functional programming, since it enables a fast prototyping and testing cycle. Table 1 presents the total number of Haskell lines used to develop the map algebra.

Table 1 – Map Algebra in Haskell

	Number of source lines		
	operations	axioms	total
Data types	6	9	15
Map Algebra	6	10	16
Auxiliary	1	5	6
Total	13	24	37

For comparison purposes, the SPRING GIS [20] includes a map algebra in the C++ language that uses about 8,000 lines of code. The SPRING map algebra provides a strict implementation of Tomlin's algebra. Our map algebra allows a more generic set of functions than Tomlin's at less than 1% of the code lines. This large difference comes from the use of the parameterized types, overloading and higher order functions, which are features of the Haskell language. Our work points out that integrating functional languages with spatial database is an efficient alternative in for developing and prototyping novel ideas in GIScience.

References

1. Frank, A. Higher order functions necessary for spatial theory development. in Auto-Carto 13. 1997. Seattle, WA: ACSM/ASPRS.
2. Frank, A. and W. Kuhn, Specifying Open GIS with Functional Languages, in Advances in Spatial Databases—4th International Symposium, SSD '95, Portland, ME, M. Egenhofer and J. Herring, Editors. 1995, Springer-Verlag: Berlin. p. 184-195.
3. Frank, A. One Step up the Abstraction Ladder: Combining Algebras - From Functional Pieces to a Whole. in COSIT - Conference on Spatial Information Theory. 1999: Springer-Verlag.
4. Medak, D., Lifestyles - a new Paradigm in Spatio-Temporal Databases, in Department for Geoinformation. 1999, Technical University of Vienna: Vienna.
5. Winter, S. and S. Nittel, Formal information modelling for standardisation in the spatial domain. International Journal of Geographical Information Science, 2003. **17**: p. 721--741.

6. Hudak, P., Conception, evolution, and application of functional programming languages. ACM Comput. Surv., 1989. **21**(3): p. 359-411.

7. Jones, S.P., Haskell 98 Language and Libraries The Revised Report. 2002.

8. Peyton Jones, S., J. Hughes, and L. Augustsson. Haskell 98: A Non-strict, Purely Functional Language. 1999 [cited; Available from: http://www.haskell.org/onlinereport/.

9. Thompson, S., Haskell:The Craft of Functional Programming. 1999, Harlow, England: Pearson Education.

10. Chakravarty, M., The Haskell 98 foreign function interface 1.0: An addendum to the Haskell 98 report. 2003.

11. Wadler, P., Comprehending monads, in Proceedings of the 1990 ACM conference on LISP and functional programming 1990, ACM Press: Nice, France. p. 61-78.

12. Jones, S.P., Tackling the Awkward Squad: monadic input/output, concurrency, exceptions, and foreign-language calls in Haskell. 2005.

13. Vinhas, L. and K.R. Ferreira, Descrição da TerraLib, in Bancos de Dados Geográficos, M. Casanova, et al., Editors. 2005, MundoGeo Editora: Curitiba. p. 397-439.

14. Chakravarty, A.P.a.M. **Interfacing Haskell with Object-Oriented Language**. in 15th International Workshop on the Implementation of Functional Languages. 2004. Lübeck, Germany: Springer-Verlag.

15. Shields, M. and S.L.P. Jones, Object-Oriented Style Overloading for Haskell. Electronic Notes in Theoretical Computer Science, 2001. **59**(1).

16. Tomlin, C.D., A Map Algebra, in Harvard Computer Graphics Conference. 1983: Cambridge, MA.

17. Câmara, G., Representação computacional de dados geográficos, in Bancos de Dados Geográficos, M. Casanova, et al., Editors. 2005, MundoGeo Editora: Curitiba. p. 11-52.

18. OGC. Open GIS Consortium. Topic 6: the coverage type and its subtypes. 2000 [cited 2006 10/05/2006]; Available from: http://portal.opengeospatial.org/files/?artifact_id=7198.

19. Câmara, G., et al. Towards a generalized map algebra: principles and data types. in VII Workshop Brasileiro de Geoinformática. 2005. Campos do Jordão: SBC.

20. Câmara, G., et al., SPRING: Integrating Remote Sensing and GIS with Object-Oriented Data Modelling. Computers and Graphics, 1996. **15**(6): p. 13-22.

An Algorithm and Implementation for GeoOntologies Alignment

Guillermo Nudelman Hess[1,2], Cirano Iochpe[1,3], Silvana Castano[2]

[1] Instituto de Informática
Universidade Federal do Rio Grande do Sul
[2] Università degli Studi di Milano
Dipartimento di Informatica e Comunicazione
[3] Empresa de Tecnologia da Informação e Comunicação de Porto Alegre

1 Introduction

The high cost in acquiring data for populating Geographic Information Systems (GIS) was, in the past, one of the main obstacles for its wide diffusion. The Internet created a huge network where users, institutions and organizations can easily share information. However, if on one hand, this interchange offers a lot of benefits, on the other hand it generates the need to address the heterogeneities among the information obtained from distinct sources.

One way of making the information's meaning more explicit is the use of ontologies [22], also in the geographic field. Some initiatives for creating geographic ontologies are described in [1,4] and by the ISO 19109 standard as well.

As the ontologies may be created by different communities, heterogeneity problems may arise when integrating the information from two or more ontologies. A number of works and tools address the problem of semantic integration (or alignment) for conventional, non geographic, ontologies [3, 6, 10, 17]. However, as they are not conceived for dealing with the peculiar characteristics of the GIS data, such as spatial relationships [14, 20], very often using conventional ontology alignment tools for geographic data does not achieve as good results as the ones obtained with conventional information. Basically, the alignment of geographic information must address the same issues addressed in conventional integration and, in addition, must consider the spatial relationships, the geometries and, in case of instances, the geographic location (coordinates) information.

In this paper we present the G-Match, an algorithm and an implementation of a geographic ontology matcher. Taking as input two different geographic ontologies, G-Match measures the similarity of their concepts by considering their names, attributes, taxonomies, conventional relationships and topological relationships. Except for the name comparison, G-Match considers both the commonalities and the differences for evaluating the similarity between two concepts.

The paper is organized as follows. Related work regarding geographic information alignment and integration is outlined in Section 2. A motivating example is presented in Section 3. Section 4 provides the G-Match algorithm description while some experimentation results of its execution are discussed in Section 5. Finally, conclusions and future work are discussed in Section 6.

2 Related work

In this section, we describe related work on geographic ontology alignment and integration by classifying the proposed approaches/techniques in three categories: ontology mediated, semantic annotation-based and spatial relationship-based.

2.1 Ontology mediated alignment/integration

Rodriguez, Egenhofer and Rugg [19] propose an approach for assessing similarities among geospatial feature class definitions. The similarity evaluation is basically done over the semantic interrelation among classes. In particular, not only the IS-A (taxonomic) and part-Of relations are considered, but also the distinguish features (parts, functions and attributes) [18]. In addition to semantic relations and distinguish features, two more linguistic concepts are taken into consideration for the definition of entity classes: words and meanings, and synonymy and polysemy (homonymy). Using ontologies and the set theory properties and operations, they determine semantic similarity among entity classes from different ontologies [18], but not considering the geospatial classes.

Fonseca et al. [8] propose an ontology-driven GIS architecture to enable geographic information integration. In this proposal, the ontology acts as a model-independent system integrator [8]. The work of Fonseca et al. [9] focuses on the application level, in which they can work on the translation of a conceptual schema into an application ontology. A framework for mapping management defines the mappings between all the elements of a spatial ontology and the constructors of a conceptual schema for geographic information.

Hakimpour and Timpf [12] propose the use of an ontology in the resolution of semantic heterogeneities, especially those found in Geographic Information Systems. The goal is to establish equivalences between conceptual schemas or local ontologies. Basically the process is done in two phases [11]. First, a reasoning system is used to merge formal ontologies. The result of merging is used by a schema integrator to build a global schema from local schemas. Then, the possible meaningful mappings in the generated global schema are found, used to establish the data mappings between the databases. Then the data (instances) from the local schemas are mapped. This process is composed by three parts: entity mapping, attribute mapping, data transformation [12].

Sotnykova et al. [21] state that the integration of spatio-temporal information (first schema and then data) is a three-step process comprising pre integration (resolution of syntactic conflicts), Inter-Schema Correspondence Assertions (ICAs) (resolution of semantic conflicts) and integrated schema generation (resolution of structural conflicts). For semantic conflicts they propose an integration language, which allows the specification of correspondences between different database schemas. Part of the work concerns how to integrate the schemas, not from the point of view of rule formalization or measures of similarities, but rather in terms of how much information the integrated schema should possess.

Stoimenov and Djordjevic-Kajan [24] propose the GeoNis framework to enable the semantic GIS data interoperability. It is based on mediators, wrappers and ontologies. In this work ontologies act as a knowledge base to solve semantic conflicts such as homonyms, synonyms and taxonomic heterogeneities.

Matching complex geographical objects based on the matching of their child objects is the proposal of Cruz et al. [5]. They have designed and implemented a tool for aligning ontologies by proposing a semi-automatic method for propagating the mappings along the ontologies, especially those for land use. In this approach, there must be a global ontology that is the reference for the alignment of the local ontologies, namely for the identification of semantically related entities in different ontologies. The alignment process is semi-automatic, in that the values associated with the vertices can be assigned in two ways: as functions of the children vertices or of the user input. The user initially identifies the hierarchy levels in the two ontologies to be aligned. Then the alignment component propagates the mappings to the parent nodes.

2.2 Semantic annotation-based alignment/integration

The Knowledge and Information Management (KIM) platform provides an infrastructure and related services for automatic semantic annotation, indexing, and retrieval of unstructured and semi-structure content. The ontologies and knowledge bases are kept in repositories based on cutting edge Semantic Web technology and standards including RDF repositories, ontology middleware and reasoning [15]. The main idea behind KIM is the semantic annotation, which means that the system looks at the description of an entity searching for keywords and then associates them with a concept in the ontology (central knowledge base). The spatial features of a concept are described in the KIMO's sub ontology. The goal is to include the most important and frequently used types of location, including relations between them (such as hasCapital, subRegionOf), as well as relations between locations and other entities and various attributes.

2.3 Spatial relationship-based alignment/integration

Focus on the semantic relationships other than the taxonomic ones is the proposal of [13]. Especially, the so called functional relations of concepts are of interest, and are available in the glosses (descriptions) of the concepts. Doing that, it is possible to find that two concepts are semantically related even though they are hierarchically not. The evaluation of concepts similarity exploits conceptual regions: the concepts are represented as n-dimensional regions in a vector space, that is, the region is continuous, completely closed and the hull of the region is convex. The measurement of semantic similarity between conceptual regions is based on applying distance measures defined in [20].

Detecting similarities between geospatial data considering different geometries is proposed by Belussi et al. [2]. In this work, the authors make a deep comparison regarding topological relationships by defining equivalences among this kind of spatial relationships depending on the geometry of the involved objects.

With respect to these works, the original contribution of G-Match regards the fact it is conceived to encompass into a unique matcher various features that may have influence on the similarity measurement when comparing two geographic ontologies. The existing proposals, in general, consider only a subset of these features, but not all of them together.

3 Motivating example

Let us consider the ontology alignment scenario below, where the ontology in Figure 1 has to be matched with the ontology of Figure 2. This example is quite simple, but complete in terms of the geographic elements usually found in geographic ontologies, which are important to describe the matching techniques we are going to present.

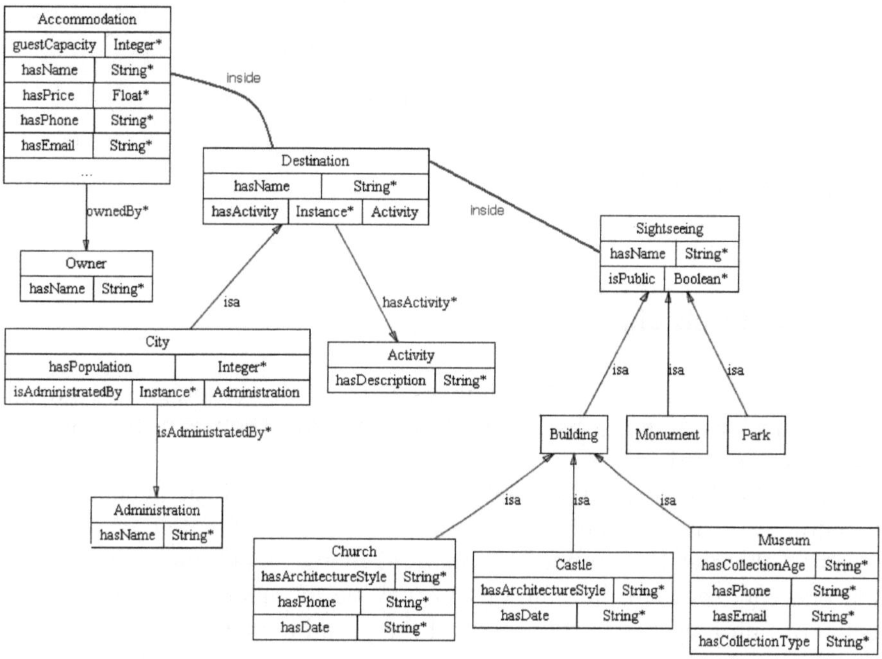

Fig. 1. The ontology *O*

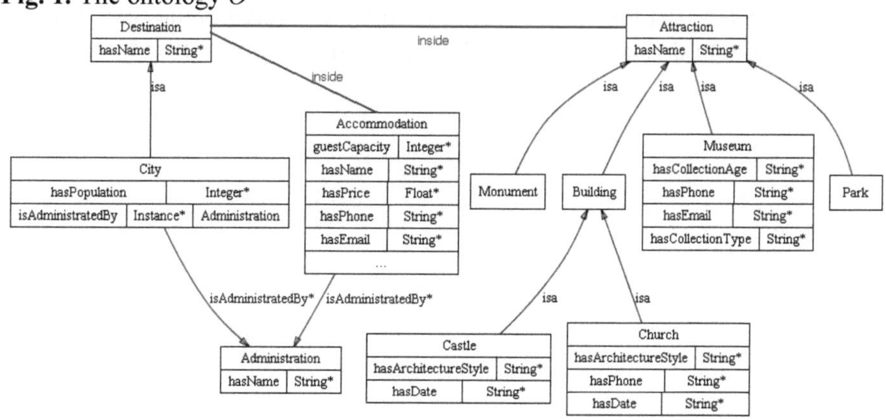

Fig. 2. The ontology *O'*

Furthermore, a specific problem of geographic ontologies, not supported by conventional matchers, occurs when the concepts to be compared are designed using different resolutions. In this case different topological relationships may have the same semantics, as stated in [2] and illustrated in Figure 3.

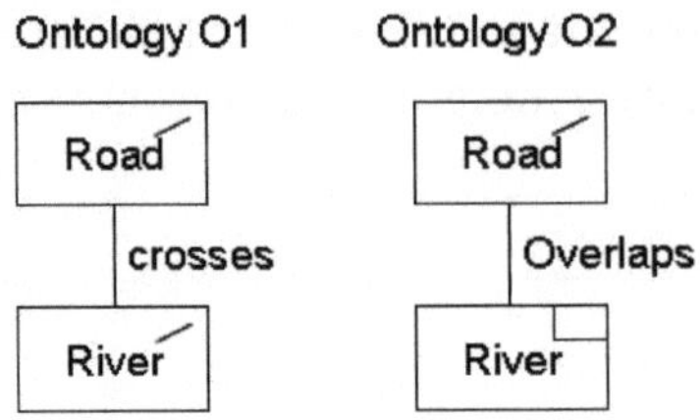

Fig. 3. Equivalent topologies

4 The G-Match algorithm

As shown in Figure 4, G-Match takes as input two ontologies *O* and *O'* and produces as output a list of similarity measures between the concepts of the two ontologies.

The G-Match algorithm is iterative, which means that each concept c_i of an ontology *O* is compared against all concepts c_j of ontology *O'*. Furthermore, the matching process is many-to-many, which means that more than one concept c_j of the ontology *O* may be the match for a given concept c_i of ontology *O'*. In this case, all the possible matching concepts are presented to the user.

The WordNet [16] lexical system is used by G-Match name matching to find synonyms and related terms, when comparing names of concepts and attributes as well.

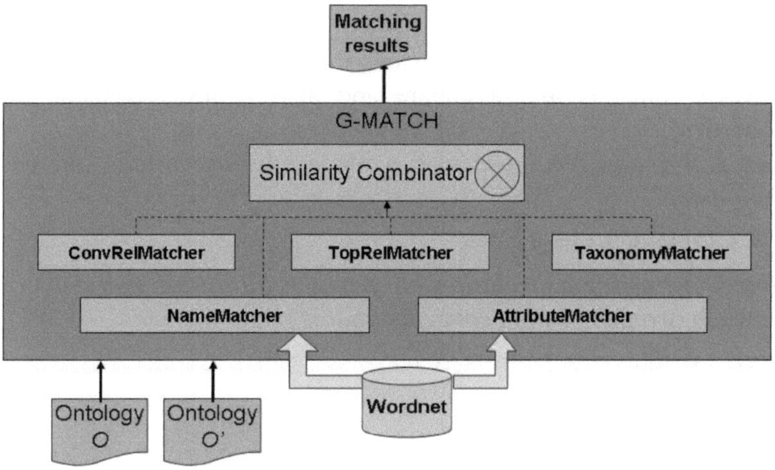

Fig. 4. The G-Match architecture

4.1 Basic definitions

A geographic ontology may contain both geographic and non-geographic (conventional) concepts. Moreover it describes the properties of a concept and the relations it has with the other concepts.

Definition 1. *A geographic ontology O is a 4-tuple of the form O=(C,I,P,X)*, where:

- C is the set of concepts. A concept $c \in C$ is identified by the synonym set T of terms which nominates it;
- *I* is the set of instances;
- *P* is the set of properties, which is defined as a tuple P = (A,R) is the set of properties of the concept c.
  - o A is the set of attributes associated with c. An attribute $a \in A$ is a binary relation of type a(c,dtp), where dtp is a data type (e.g., string, integer, etc.)
  - o *R* is the set of relationships of c with other concepts. A relationship $r \in R$ is a binary relation *r(c,c')*. Furthermore, a relationship *r* = *{g,tr,cr}*, where *g* is a relationship between the concept *c* and a concept *c'* which denotes a geometry, *tr* is a topological relationship, i.e., a special type of spatial relationship between two geospatial concepts *c*

and *c'* and *cr* is a conventional relationship between two concepts *c* and *c'*;

- *X* is set of axioms, which describes the hierarchical relationships between concepts and associate and also restricts concepts to properties.

Definition 2. *A geographic concept is a concept having at least one geometry relationship,*

$$gc = \{c \in C \mid \exists r(c,c'), c' \in g\}$$

Definition 3. *A topological relationship tr is a relationship that can occur only between two geographic concepts, that is*

$$tr = \{r(c,c') \in R \mid (c = gc) \wedge (c' = gc)\}$$

4.2 The Algorithm

G-Match performs similarity evaluation in three main phases, as shown in Figure 5. In the first phase, concepts names (*SimName*(c_i,c_j)) and attributes (*SimAt*(c_i,c_j)) are matched. Then, the second phase of similarity evaluation is executed, which considers taxonomies (*SimTx*(c_i,c_j)), conventional relationships (*SimRel*(c_i,c_j)) and topological relationships (*SimTop*(c_i,c_j)). The last phase calculates the overall similarity value by combining the results of the previous phases. The algorithm works as follows.

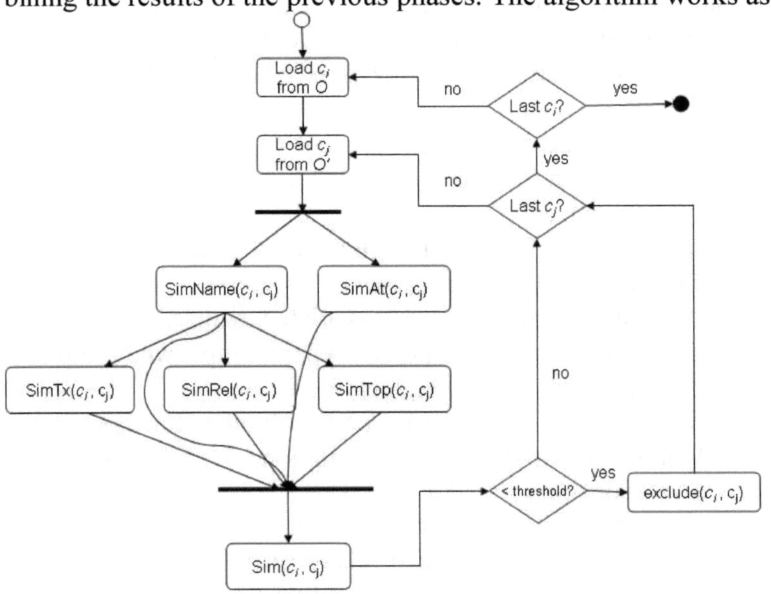

Fig. 5. G-Match execution flow

1. **Load concept c_i from ontology *O***: A concept c_i is loaded.
2. **Load concept c_j from ontology *O'***: A concept c_j from the other ontology is loaded to be compared against the concept c_i.
3. **Measure the name similarity of c_i and c_j**: using the WordNet tesaurus as an auxiliary lexical base, the name similarity $SimName(c_i,c_j)$ is calculated by searching the correspondence of the two terms $t(c_i)$ and $t(c_j)$. If WordNet returns 0 (i.e., no terminological relationships are retrieved), we calculate the string similarity of the terms using a combination of the edit distance and Jaro-Winkler metric as proposed in [23].
4. **Measure the attribute similarity of c_i and c_j**: we use again WordNet to assess the similarity between the names of the attributes. We consider as matching attributes only those whose names coincide or are synonyms. For attribute matching, every attribute $a(c_i,dtp) \in A(c_i)$ is compared against every attribute $a(c_j,dtp) \in A(c_j)$. The final similarity measure regarding the attributes is given by:

$$SimAt(c_i,c_j) = \frac{\sum((a(c_i) \cap a(c_j)) * W_a)}{A(c_i) \cup A(c_j)} \tag{1}$$

where W_a is the weight of the attribute a in the ontology. This weight is inversely proportional to the number of concepts having the attribute. The more concepts, the more generic the attribute is and thus the lower its weight is. The steps 3 and 4 can be executed in parallel.

5. **Measure the hierarchy similarity of c_i and c_j.:** based on the results obtained in the step 3, the similarity based on the concept taxonomy is measured. This is done by checking the number of common sub-concepts of concepts c_i and c_j and by considering their hierarchy level. The final value for the taxonomy similarity measure is given by:

$$SimTx(c_i,c_j) = \frac{\sum((h(c_i) \cap h(c_j)) * W_{level})}{H(c_i) \cup H(c_j)} \tag{2}$$

where W_{level} is 1.0 if the subclasses are at the same level and it is decreased by an arbitrary value if they are at different hierarchical levels in the two ontologies.

6. **Measure the relationship similarity of c_i and c_j**: using the results from step 3, the similarity of the conventional relationships is

measured. This is done by counting the common relationships the two concepts c_i and c_j have in common over the total number of relationships, as follows:

$$SimRel(c_i, c_j) = \frac{cr(c_i, c_i^{'}) \cap cr(c_j, c_j^{'})}{cr(c_i, c_i^{'}) \cup cr(c_j, c_j^{'})} \tag{3}$$

7. **Measure the topological relationship similarity of c_i and c_j**: in G-Match, using the results from step 3, the similarity of the topological relationships is measured. We consider the topological relationships described in [7], namely *disjoint*, *touch*, *inside*, *cover*, *coveredBy*, *overlap*, *equal*, *cross* and *contain*. The similarity value is given by:

$$SimTop(c_i, c_j) = \frac{tr(c_i, c_i^{'}) \cap tr(c_j, c_j^{'})}{tr(c_i, c_i^{'}) \cup tr(c_j, c_j^{'})} \tag{4}$$

For the topological relationships, it is important to clarify that we do not consider only the name of the relationship, but also the geometries of the concepts. As described in [2], depending on the geometries of the concepts, the different topological relationship have the same meaning, that is, are equivalent. The G-Match is capable of detecting these equivalences during the similarity measurement, and this is the main feature that makes it suitable for geographic ontologies.

8. **Measure the overall similarity of c_i and c_j**: in this step, the similarity values obtained in the previous steps are combined in a comprehensive value through a weighed sum, as follows:

$$\begin{aligned} Sim(c_i, c_j) = {} & WN * SimName(c_i, c_j) + WA * SimAt(c_i, c_j) + \\ & WH * SimTx(c_i, c_j) + WR * SimRel(c_i, c_j) + WT * SimTop(c_i, c_j) \end{aligned} \tag{5}$$

where WN is the name similarity weight, WA is the attributes similarity weight, WH is the hierarchy similarity weight, WR is the relationships similarity weight and WT weight the topological similarity weight. The sum of the weights is equal to 1 and thus the overall similarity measure value is within [0,1].

9. **Discard non-relevant matches**: the pairs (c_i,c_j) with low similarity values are automatically discarded, in order to produce significant results and make it easier to choose the correct matches. Thus, a threshold-based is adopted and a threshold parameter value is properly set at the beginning of G-Match execution.
10. **Iterate concept matching in *O'***: if there are more concepts from ontology *O'* to be processed, return to step 2.

11. **Iterate concept matching in *O*:** if there are more concepts from ontology *O* to be processed, return to step 1.

5 Experimental results

In order to experiment G-Match, we executed the algorithm using as input the geographic ontologies presented in the Section 2, which have the following main differences:

- The whole hierarchy of *TransportFacility* is present only in the ontology *O*;
- The concept *Attraction* of ontology *O* is equivalent to the concept *Sightseen* in ontology *O'*; the most similar concept to *Hotel* of ontology *O* in ontology *O'* is *RegularHotel*;
- Accommodation in ontology *O* is associated with *Administration*, while in ontology *O'* is has a relationship with *Owner*;
- For some concepts, some attributes are present only in ontology *O*;

We implemented the G-Match in two ways: as a stand-alone, complete matcher (called G-Match complete) and as an extension for an existing conventional matcher, H-Match [3] (called G-Match). In this latter case, only the relationship (conventional and topological) similarity was measured by our tool. For experimentation, we selected two generic ontology matching tools, namely Prompt [17] and H-Match [3]. Here we used them to match the geographic ontologies *O* and *O'* of Section 2 with the goal of analyzing and comparing the results obtained with such generic matchers and with our specific geographic matcher G-Match.

The tests have been performed using a threshold value of 0.4. Table 1 shows the results in terms of recall and precision. EM denotes the expected matches, AM the matches detected automatically (i.e., similarity measures higher than 0.7) and CAM the correct matches automatically detected. The thresholds value have been selected because they offered the best results, after trying several combinations.

Table 1. Precision and recall results

Matcher	EM	AM	CAM	Precision	Recall
Prompt	15	12	10	83%	67%
H-Match	15	12	11	92%	73%
G-Match	15	13	12	92%	80%
G-Match Complete	15	15	14	93%	93%

As it can be seen, using a matcher specially tailored for the spatial relationships increases both the recall and precision. Furthermore, in the cases where G-Match failed in choosing the correct match there were more than one pair(c_i, c_j)with similarity value higher than 0.7. The expected correct match was one of the returned pairs, but not the one with higher similarity. When the G-Match did not find any pair(c_i, c_j) with similarity higher than the matching threshold, the correct pair was within the ones with similarity higher than the threshold.

Conclusions and future work

The challenge faced here was to develop a methodology and techniques to achieve good results when matching two geographic ontologies. The G-Match matching process is balanced, that is, it considers the features of a concept separately and then weights each feature (name, attributes, taxonomy, conventional and topological relationships) to compute the overall similarity between two concepts. The initial experimental results obtained show that the G-Match is in the correct direction towards the development of a semantic matcher specially tailored for geographic ontologies.

As future work, we plan to study the impact of the other spatial relationships, such as distance relations, on the similarity evaluation of ontology concepts. Furthermore, up to now the G-Match always considers all the features, independently of the concept being processed. Thus, if the ontology does not have, for example, topological relationships, the similarity measure decreases. Because of that, we intend to make the G-Match capable of self-adaptation depending on the input ontology, which means self-configuration of the weights WN, WA, WH, WR and WT based on the structural characteristics of the input ontologies. At last, the alignment of geographic ontologies at the instance level is also planned as future work.

Acknowledgment

The authors thank the Brazilian agency CAPES (Coordenação de Aperfeiçoamento de Pessoal de Nível Superior) for the financial support.

References

[1] Apinar, I. B., Sheth, A., Ramakrishnan, C., Usery, E. L., Azami, M., and Kwan, M.-P. (2005). Geospatial ontology development and semantic analysis.

InWilson, J.P. and Fotheringham, S., editors, Handbook of Geographic Information Science. Blackwell Publishing.

[2] Belussi, A., Catania, B., and Podestà, P. (2005). Towards topological consistency and similarity of multiresolution geographical maps. In GIS'05: Proceedings of the 13th annual ACM international workshop on Geographic information systems, pages 220– 229,NewYork,NY, USA.ACM Press.

[3] Castano, S., Ferrara, A., and Montanelli, S. (2006). Matching ontologies in open networked systems: Techniques and applications. In Spaccapietra, S., Atzeni, P., Chu, W.W., Catarci, T., and Sycara, K.P., editors, J. Data Semantics V, Lecture Notes in Computer Science, pages 25–63. Springer.

[4] Chaves, M. S., Silva, M. J., and Martins, B. (2005). A geographic knowledge base for semantic web applications. In Heuser, C. A., editor, SBBD, pages 40–54. UFU.

[5] Cruz, I.F., Sunna, W., and Chaudhry, A. (2004). Semi-automatic ontology alignment for geospatial data integration. In Egenhofer, M. J., Freksa, C., and Miller, H. J., editors, GIScience, volume 3234 of Lecture Notes in Computer Science, pages 51–66. Springer.

[6] Doan, A., Madhavan, J., Domingos, P.,and Halevy, A.Y. (2004). Ontology matching:A machine learning approach. In Staab, S. and Studer, R., editors, Handbook on Ontologies, International Handbooks on Information Systems, pages 385–404. Springer.

[7] Egenhofer, M. J. and Franzosa, R. D. (1991). Point set topological relations. *International Journal of Geographical Information Systems*, 5:161–174.

[8] Fonseca,F., Egenhofer, M., Agouris,P., and Camara, G. (2002). Using ontologies for integrated geographic information systems. Transactions in Geographic Information Systems, 6(3).

[9] Fonseca,F.T., Davis,C.A., and Camara, G. (2003). Bridging ontologies and conceptual schemas in geographic information integration. GeoInformatica, 7(4):355–378.

[10] Giunchiglia, F., Shvaiko, P., and Yatskevich, M. (2005). S-match: an algorithm and an implementation of semantic matching. In Kalfoglou, Y., Schorlemmer, W. M., Sheth, A.P., Staab, S., and Uschold, M., editors, Semantic Interoperability and Integration, volume 04391 of Dagstuhl Seminar Proceedings. IBFI, Schloss Dagstuhl, Germany.

[11] Hakimpour, F. and Geppert, A.(2002). Global schema generation using formal ontologies. In Spaccapietra, S., March, S.T., and Kambayashi, Y., editors, ER, volume 2503 of Lecture Notes in Computer Science, pages 307–321. Springer.

[12] Hakimpour, F. and Timpf, S. (2001). Using ontologies for resolution of semantic heterogeneity in GIS.

[13] Jiang, J. and Conrath, D. (1997). Semantic similarity based in corpus statistics and lexical taxonomy. In International Conference Research in Computational Linguistics, ROCLING X, Taiwan.

[14] Kuhn, W. (2002). Modeling the semantics of geographic categories through conceptual integration. In Egenhofer, M. J. and Mark, D. M., editors, GIS-

cience, volume 2478 of Lecture Notes in Computer Science, pages 108–118. Springer.

[15] Manov, D., Kiryakov, A., Popov, B., Bontcheva, K., and and, D. M. (2003). Experiments with geographic knowledge for information extraction. In Geographic References, HLT/NAACL'03, Edmonton, Canada.

[16] Miller,G.A. (1995).Wordnet:Alexical databasefor english. *Commun.ACM*,38(11):39–41.

[17] Noy,N.F. (2004).Toolsfor mapping and merging ontologies. In Staab,S. and Studer, R., editors, Handbook on Ontologies, International Handbooks on Information Systems, pages 365–384. Springer.

[18] Rodriguez, M. A. and Egenhofer, M. J. (2003). Determining semantic similarity among entity classes from different ontologies. IEEE Trans. Knowl. Data Eng., 15(2):442– 456.

[19] Rodriguez, M. A., Egenhofer, M. J., and Rugg, R. D. (1999). Assessing semantic similarities among geospatial feature class definitions. In Vckovski, A., Brassel, K. E., and Schek, H.-J., editors, INTEROP, volume 1580 of Lecture Notes in Computer Science, pages 189–202. Springer.

[20] Schwering, A. and Raubal, M. (2005). Spatial relations for semantic similarity measurement. In Akoka, J., Liddle, S. W., Song, I.-Y., Bertolotto, M., Comyn-Wattiau, I., Cherfi, S. S.-S., vandenHeuvel,W.-J., Thalheim, B., Kolp, M., Bresciani,P., Trujillo, J., Kop, C., and Mayr, H. C., editors, ER (Workshops), volume 3770 of Lecture Notes in Computer Science, pages 259–269. Springer.

[21] Sotnykova, A., Cullot, N., and Vangenot, C. (2005). Spatio-temporal schema integration with validation: A practical approach. In Meersman, R., Tari, Z., Herrero, P., Mendez, M. S., Robles, V., Humble, J., Albani, A., Dietz, J.L.G., Panetto, H., Scannapieco, M., Halpin, T.A., Spyns, P., Zaha, J. M., Zimanyi, E., Stefanakis, E., Dillon, T. S., Feng, L., Jarrar, M., Lehmann, J., de Moor, A., Duval, E., and Aroyo, L., editors, OTM Workshops, volume 3762 of Lecture Notes in Computer Science, pages 1027–1036. Springer.

[22] Spaccapietra, S., Cullot, N.,Parent, C., and Vangenot, C. (2004). On spatial ontologies. In Proceedings of the VI Brazilian Symposium on Geoinformatica (GEOINFO 2004), Campos do Jordão, Brazil.

[23] Stoilos, G., Stamou, G., and Kollias, S. (2005). A string metric for ontology alignment. 4th International Semantic Web Conference (ISWC 2005), Galway, 2005.

[24] Stoimenov, L. and Djordjevic-Kajan, S. (2005). An architecture for interoperable GIS use in a local community environment. Computers and Geosciences, 31:211–220.

Querying a Geographic Database using an Ontology–Based Methodology

Renata Viegas[1], Valéria Soares[1,2]

[1]Programa de Pós-Graduação em Sistemas e Informação
Universidade Federal do Rio Grande do Norte
[2] Departamento de Informática
Universidade Federal da Paraíba

1 Introduction

Geographic Information Systems (GIS) are multidisciplinary systems that could be used by different community users, each one with their own objectives and interests. So, different visions of the same reality must be combined to support the community's necessities.

Different people recognize differently the same geographic region. Geographic features are collected and stored in GIS that were modeled by some specific conceptual model. If we need an efficient search on geographic databases, in some cases, it is necessary to associate meaning with the data. So, current GIS must be able to solve the semantic interoperability due to the fact that a geographic feature could have more than one description. The term semantic refers to the meaning of these features.

According to Fonseca [6], information systems must be able to understand the user's models and their meanings. In other words, we could say that it is necessary to develop systems that exceed the information barriers to give to users not only the data, but also their meaning.

For spatial data interchange happens without missing information it is necessary a high degree of interoperability between GIS's [6]. Regarding to interoperability between different GIS formats, some alternatives have been proposed to prevent this problem, as the creation of standards like

SDTS (Spatial Data Transfer Standard) [15] and SAIF (Spatial Archive and Interchange Format) [16].

Although standards for exchange of data are necessary and useful, they do not have the capacity to transfer the meaning associated with these data. Nowadays there is an increasing interest about how to reach interoperability with the use of ontologies as being a knowledge database type. The ontologies could specify a specific vocabulary domain relative, and could define entities, classes, properties, predicates, functions, and the relationships between these components.

The term ontology comes from Philosophy, meaning the representation of existence through a systematic explanation, as being the conception of all things that may "exist" or "be". In the computer science area, ontology began to be used by Artificial Intelligence, as being a "formal and explicit specification that tries, in the best way, to make the defined world structure to be closer by a concept" [8]. On this definition, "formal" means computer readable; "explicit" is concerning concepts, properties, relations, functions that are explicit defined; "concept" concerns an abstract model of some phenomenon of the real world [8].

An explicit formalization of our mental model is generally called ontology (with a lower–case "o"). The basic description of the real things of the world, the description of what would be true, is called Ontology (with an upper case "O"). Thus, there is only one Ontology, but several ontologies [6]. Each community that offers information and accesses them has his own ontology. Each one of these ontologies may be divided on small ontologies. The details level of ontologies tells the geographic information details level [6].

A geographic ontology is a conceptualization of a phenomenon or geographic object in the real world. It is necessary to store all characteristics referred to a geographic object. This is what differs from a geographic ontology and other types of ontologies. Besides the domain being geographic, characteristics of geographic objects (location, topology, direction) are embodied to this ontology. The search for geographic data semantics is important for the interoperability among GIS.

The development of ontology is an iterative process. To build ontology consists on learning and understanding the concepts and visions that are relevant for the different users of a GIS.

This aim work is to provide a mechanism to allow that different communities' users access the same Geographic Database without knowing its internal structures, and using only specific terms of each research area. Although we have only one database implementation, through the definition of different communities' ontologies, anyone could search the database, in a transparent way, using a specific interface. No one will need to know how the database is, in fact, implemented. But, everyone could search for information, and will have their queries attended. Through the relationships between the defined ontologies, different descriptions and names about the same data could be merged and implemented with appropriated mechanisms. These mechanisms use equivalent classes mapping, and an intelligent GIS layer that interact with the ontologies and with the geographic database, and that give to the user the answers about his queries, independent of the used terms.

The presented article is organized as follows: section 2 shows a gather of the related works. Section 3 explains the system architecture of our work, Section 4 present the application in a specific geographic domain, and finally Section 5 concludes this paper.

2 Related Works

Researches about interoperability in information systems are motivated by the increasing heterogeneity in computer world.

Guarino [8] proposed the use of ontologies on information systems, conducing to ontology–driven information systems – ODIS.

Heterogeneous data on GIS is not an exception, but the complexity and diversity of geographic data and the difficulty of its representation make this search for interoperability on this kind of system more complex.

Fonseca [6] defends the use of geographic ontologies for the semantic integration of data in GIS. On his work, he defines ODGIS (Ontology–Driven Geographic Information Systems), which are systems that use translated ontologies on software components.

Egenhofer report on [4] the creation of the Semantic Geospatial Web: a framework for geospatial information retrieval based on the semantics of spatial and terminological ontologies. This framework enable users to retrieve more precisely the data they need, based on the semantics associated with these data.

Others ontology–based works for query formulation could be found in [3], [7]. In [2] it is presented an architecture that binds the web semantic concepts with regular expression techniques whose objective is to recover and mine data from web pages.

An algorithm is proposed in [10] to align and merge ontologies, by name similarities between classes. In this algorithm, for each class of all the ontologies, the SMART algorithm automatically executes a series of actions, asking the users if classes with the same name could be aligned in a same ontology.

In this work, we propose to create a semantic layer between the geographic database and the users, using ontologies. With a unique database, we could have a low cost implementation because we have only one data collect, and different users, with different knowledge, could access this same database. Furthermore, we develop a case study on environmental and marine modeling, what increases the difficulty of collecting the information, and none of the proposal quoted on literature is similar to this.

3 System Architecture

We propose a solution based on geographic ontologies to provide that different professionals of distinct research areas access the same Geographic Database (GDB).

From the different terms stored on these ontologies, the system could infer necessary information to the users' queries, allowing that different community users access and interact with the system, without know specific characteristics about the internal structure of the database.

We create a semantic layer that intermediate the users' queries with the geographic database. Each one of the community users could interact with the system using only specific terms of his research area, and could receive his queries answers in an appropriate way. It is transparent to the user how this database was effectively implemented. The user only has to worry about his necessity, and what he wants to look for on the database. Through the semantic layer, the ontologies will be activated, and these users' queries will be translated to an appropriate SQL clause.

We could see the complete system architecture on Figure 1 that shows the semantic layer, between the users, the application and the database.

In this work, like we can see in the Figure 1, we developed the geographic database with the Database System PostGIS [13], developed the ontologies on the language OWL [12], with the editor Protégé [14], and use the API Jena [9] to generate the ontologies graphs. The package jena.ontology [9] has the classes *OntClass* and *OntModel* that allows that graphs components could be accessed by many ways.

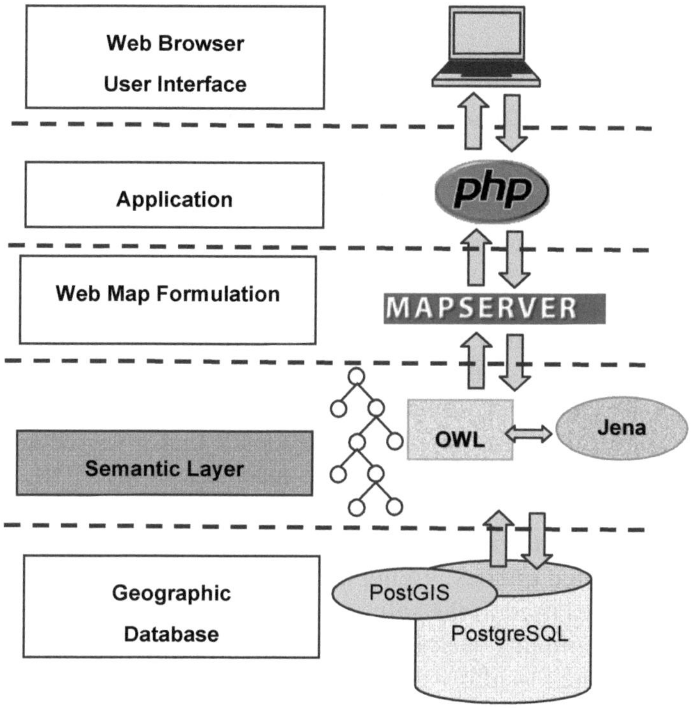

Fig. 1. System Architecture with the Semantic Layer

3.1 Geographic Ontologies

There are many definitions about ontologies in the literature [8] [6] [1].

Some of these references provide definitions for what a geographic ontology is. Sometimes a geographic ontology is defined as simply an ontology whose domain is geographic. In this work we propose to add some more characteristics to these ontologies to differentiate them to others. For this, we decided to use typical geographic relationships to relate the ontologies classes with each others, like topological, metrics and directions relationships.

We defined relationships between each ontology classes, using the properties function of the ontology specification on Protégé Editor [14]. These properties are used to spatial relate a concept to another, based with the basic operations defined by the 9–Intersection Model [5], and also specified in the OpenGIS Consortium [11]. This can be verified in the following examples, where we present some concepts that were spatially related with another, such as:

Reef Crest	*ISA* →	Reef Component
Reef Crest	*TOUCHES* →	Fore Reef
Reef Crest	*TOUCHES* →	Back Reef
Floating Boats	*SURROUND* →	Dip in the Sea Area

This spatial information in the ontologies' relationships and properties could be useful in the solution of some users' queries, as well as in the complement of others. Through pop–up windows, for each submitted query, the application also supplies a dictionary that gives detailed information (descriptive and spatial) about the instances of the GDB, that were effectively used in the solution of the queries, as well as its relationships. The objective to use these spatial relationships is to give support to users about their submitted queries, giving them additional information about the geographic features used in the solution of his queries.

Each user' community has a pay–define query interface. This specific interface was created to allow the user interact with the system using only the specific knowledge to its area of performance.

We can see another vision of the system architecture in Figure 2. The query will be processed on this following way: the user query will be submitted, and the system will recognize the used terms, and relate with the specific ontology of this type of user at this moment. So, we have to identify the user query based on the concepts and terms that had been used in this submission.

After we had identified the query, the next step is to look for the terms and concepts used in this query, in the users' ontology, comparing this ontology with the GDB ontology, looking for equivalents concepts. Through the ontologies' URLs, the Jena API [9] will be used to construct the graphs of the ontologies.

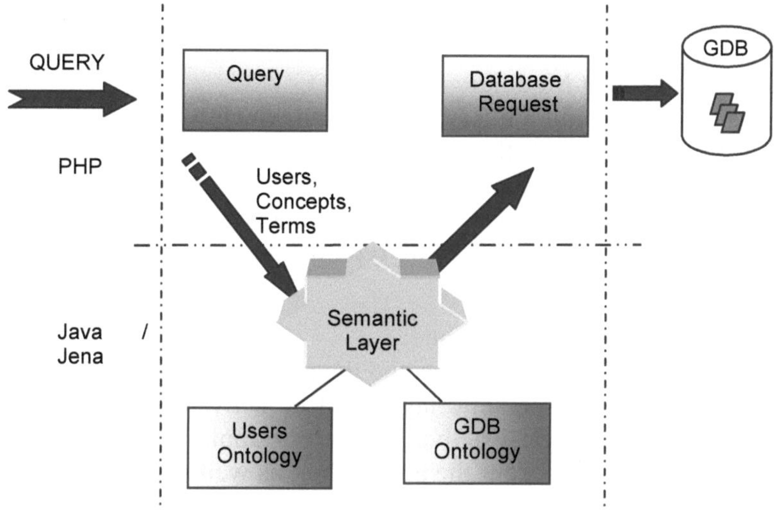

Fig. 2. General system architecture

3.2 Detailing the Semantic Layer

The next step, as well as define the ontologies, is the binding definition between the classes of the different ontologies. The result of this binding process is a formal structure with expressions that show which terms of determinate ontology is related to others terms of another ontology.

We have done a mapping with synonyms classes, to help the localization of classes whose information is relevant to answer the user queries. To compare and join the ontologies, in this work we do not use classes that have the same name in distinct ontologies. To compare and join the ontologies, in this work we use not only classes that have the same name in distinct ontologies, but also classes with different names, too. We consider synonym classes those whose concepts have the same meaning, independent of their given names, that is related of the specific knowledge of each community.

After we have done this mapping on the defined ontologies, we have also to manipulate them. So, it is necessary to go along the terms of each ontology looking for similar concepts. We use the Jena API [9] to generate the ontologies graphs. It is possible to generate graphs RDF, which is represented by resources, properties and literals. From the Jena API methods we can manipulate and compare the ontologies.

A detailing of the semantic layer could be seen in Figure 3. The layers' modules are:

- Users' Management Module:

The first step to submit a query is to inform which type of user wants to interact with the system. With this module, the user will be able to choose which type of user interface he wants to interact with the system. Depending on the user' choice, the system will shown a pay–define queries' interface, with only specific terms of this type of users, based on the defined ontologies.

- Ontologies' Management Module:

With the choice of the user's type and the pay–defined query's interface, the next module of the semantic layer will be activated, and will be responsible for the ontologies' activation. In our application, the activated ontologies will always be from the active user, as well as the ontology that represents the contents of the GDB. The ontologies are stored in ontologies' server, and are accessed through its URLs.

At the moment of the activation of the specific ontologies, some methods of the API Jena will be used to construct the graphs (models) of the ontologies.

The API Jena has object classes that represent graphs, resources, properties and literals. A graph is called model, and is represented by the Model interface. These models will keep the activated ontologies for the application. With the ontologies represented in graphs' form, we can make sweepings, looking for the desired terms.

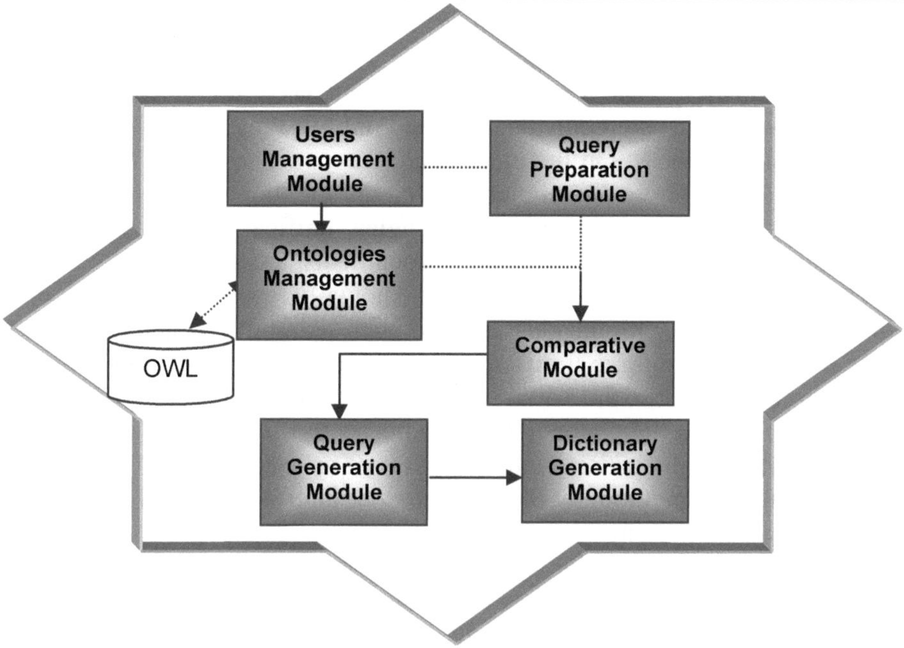

Fig. 3. Semantic Layer Modules

- Query Preparation Module:

With the pay–defined interface, the user will formulate his query, choosing what he desires to search. After that, the Query Preparation Module will be activated. This module will identify the key terms of the query.

- Comparative Module

The ontologies' construction is an iterative process. To accomplish this it is necessary to realize some interviews with professionals of the domain area to collect important terms and the relationships between them, and construct and model the ontologies.

With the ontologies created and modeled, the next step of our methodology is to continue discovering similar terms in the ontologies. In this work, the ontologies will be compared, searching similar terms. This similarity is defined manually, based on the interviews with professionals of the area. In accordance with the point of view of each professional, classes with the same meaning can be nominated with different terms.

One of the most important modules of our application is the Comparative Module. It is responsible by the search of similar terms in the ontologies. With the graphs (models) of the available ontologies, the search will be for classes that have similar concepts to the user ´s term.

The linking between the equivalents concepts is represented in OWL language through tags equivalentClass and sameAs, however has a difference between these tags. The <owl: equivalentClass> tag is used to indicate that two classes are equivalents if, and only if, they possess, necessarily, the same instances. On the other hand, the <owl: sameAs> tag is used when we have different nomenclatures that are mentioned to a same class. A typical use of *owl:sameAs* is for ontologies' unification, to say that two individuals classes, defined in different documents, are equals [12].

Thus, we use the <owl: sameAs> tag. Manually, each class of each ontology is compared with the classes of another, using the related terms that have the same meaning, but different nomenclatures. For each term found with this characteristic, the <owl: sameAs> tag is applied.

The next step to the Comparative Module is to use methods of the API Jena to treat the similarities, binding classes of an ontology to another one. The getSameAs() method, by the *OntResource* interface, is used to find the similar classes in the ontologies.

This method looks for the class passed as query parameter, in the user ontology, and search to the terms in the database ontology that have some *owl:sameAs* tag, and that the similarity is accurately the term that passed as parameter. This method also looks for in the ontologies OWL documents to the tag <owl: sameAs> that are present in the activated ontologies for the application. If this tag is found, the conceptual similarity is established.

- Query Generation Module

The Query Generation Module will mount the query that will be submitted to the database. For this, will be used the terms found in the search for the similarity, as well as the relationship used in the query interface. In next section we give an example of how the system generates the query clause.

- Dictionary Generation Module

The query will be returned for the user, locating in the map where he searched, beyond an explain window, showing all the information on its research. The Dictionary Generation Module will mount a detailed text, with the key terms of the query, supplying to the user descriptions about the geographic features involved in his queries.

4 Application: Coral Reef Domain

The geographic domain of our application is the coral reefs. We choose this domain because this work is part of a research project of the UFRN that involves researchers of distinct areas on the region of Maracajau reefs situated on the north littoral of the RN state.

In this project we have basically three different communities: the geologists, the biologists and the tourists. Because of that, we developed three different ontologies for each one of these communities.

The main objective of this work primarily was to model the mental worlds of each one of these communities by using ontologies. We choose a natural environment of coral reefs because there were not any works in literature about that use semantic terms with geographic databases.

Besides, we think that all these different communities must have the right to access the information stored on this database. So, we developed ontologies that will give support to the construction of adaptable interfaces for each community involved in the Project. This was done because we know that, although there is only one reality of coral reefs in the region of *Maracajaú*, each one of these communities think differently about that.

According to Fonseca [6], if we have a body of water, for example, for a biologist it could be a habitat for some fish, but for a firefighter it could be an emergency source of water. So, it depends on the point of view of each person, and his necessity at this time.

The first ontology that we have developed was the geologist one. This is presented on Figure 4, and was named by O_{geo}. We have to say that this ontology is much bigger then we present here, but we had to summarize to put in this paper.

On the ontology showed in Figure 4, we could see the benthic region. This region could be defined as the deepest layer of a body of water, like a lagoon, a river or the ocean. We could simply say that the benthic region is the minerals and organisms that compound the bottom of reefs.

We developed also ontology for the biologist's community and named it by O_{bio}, and ontology for the tourists community, that we call O_{tur}. Through the semantic layer that we present in the last section, we defined some equivalent classes between these ontologies.

We also have developed different interfaces for each one of these communities. So, if a biologist wants to interact with the system, he will submit a query using only specific biologist terms, without worrying about the database contents.

Next section we could show how users can access the geographic database, submitting their queries through the appropriate interfaces, and which kind of results it is generated.

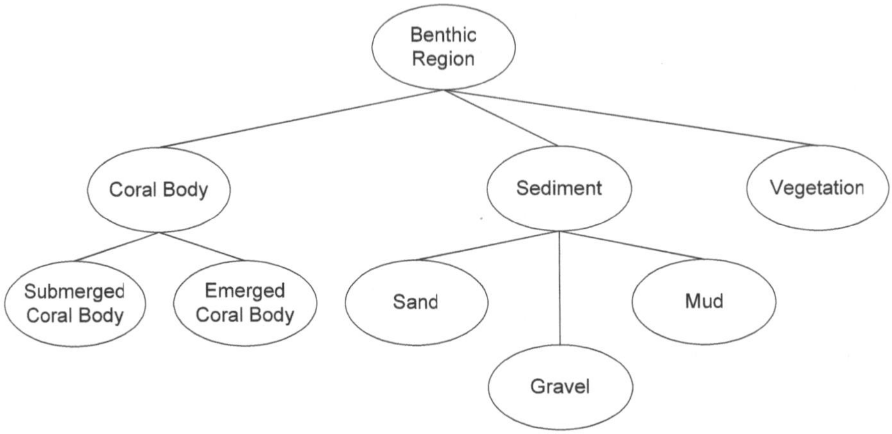

Fig. 4. Geographic Ontology for the Coral Reefs Domain

4.1 Prototype Query Examples

We have already developed all the architecture modules presented on section 3, and, based on the coral reef application ontologies we present now two different examples of query submissions.

a) A Tourist Query Submission

Suppose that a tourist wants to find the best area for dip in the sea nearby the coral reefs. This information is not stored on the database. On the other hand, we could find this information using the developed semantic layer that defined equivalences between the classes of the specific ontologies.

The activated ontologies for this query are the geologist (O_{geo}) and the tourist (O_{tur}). So, the first step to solve this user query is to look for the term dip in the sea on the tourist ontology. We defined in the O_{tur} that a tourist could dip in the sea around the floating boats on within the natural pools. The floating boats are some boats whose geographic position do not change, and is used as a base point for tourists and researchers that work in the area. After this term was found on the tourist ontology, the system will look for an equivalent term on the geologist ontology.

No similar class to "dip in the sea" could be found in the O_{geo} at this point. So, we have to go down one more step on the geologist ontology, to verify if there is some more information in the relationships between classes or subclasses.

On the other hand, going along the O_{tur} ontology, we found the relations "Dip in the sea in Natural pools" and "Dip in the sea near by floating boats". That is, the term dip in the sea appears in the tourist ontology with properties that bind this class to others classes in the ontology. The property "within" binds the class "dip in the sea" (domain) to the class "natural pool" (range). And the property "surround", binds the class "dip in the sea" (domain) to the class "floating boats" (range).

Then, the system will search now for similar classes to "natural pool" and "floating boats" on the geologist ontology. The class "floating boats" is defined both on the tourist ontology O_{tur}, and on the geologist ontology O_{geo}.

The problem is the term "natural pool" found in the O_{tur} that we still have to search for conceptual similarity on the O_{geo}.

No conceptual similarity is found in the O_{geo}, and thus the Comparative Module goes down one more step in the graph generated from the tourist ontology, looking for relationships between classes. Finally at this point, the relationship is found: "Natural pools" surround "Coral Reef";

"Coral Reef" is situated inside of "Coral Region" on the Geologist Ontology. So, we can infer that for each coral reef body that appears in the coral region, we could find natural pools around it. The information about "Coral Reef Bodies" is stored in the Geographic Database.

So, the query that will be submitted to the geographic database, after we had inferred between the two ontologies is: "show all the areas that surround the floating boats and the coral reefs bodies." And then, the generated query clause is:

```
SELECT buffer(flutuante.flutuante_geom, 10)
AS flutuante_geom,
buffer(corpo_coralineo.geom_cabeco,10)
AS  geom_cabeco
FROM flutuante, corpo_coralineo
AS foo USING UNIQUE oid USING SRID = -1
```

The operator buffer returns the area situated surrounded some geometry, based on a radius value that is passed as a parameter. In this case, we use a radius value equal to 10 meters. The result of this query is shown on Figure 5.

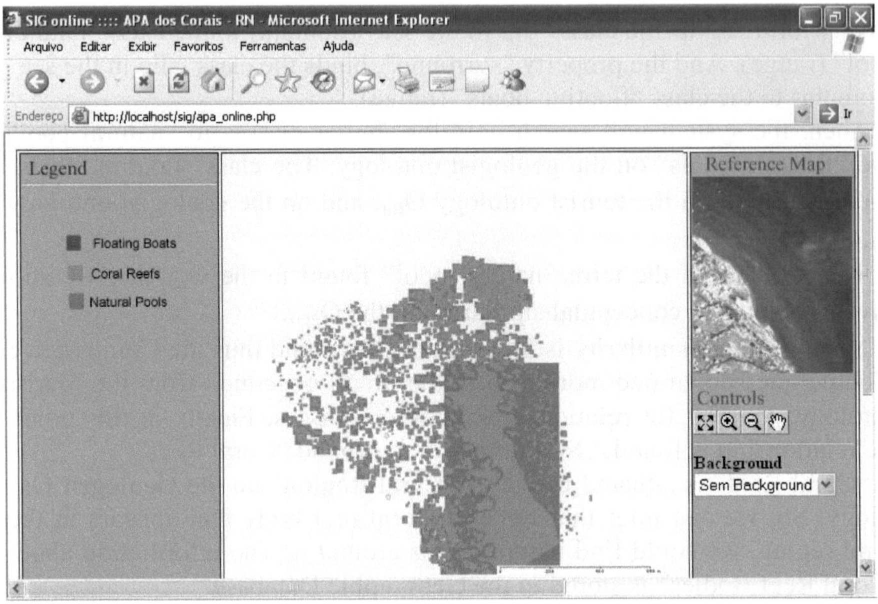

Fig. 5. Result set of the submitted query

Finally, using the information stored on the ontologies, and used to generate the queries, the system can show detailed information about the geographic features that were used on the queries, like we can see on Figure 6.

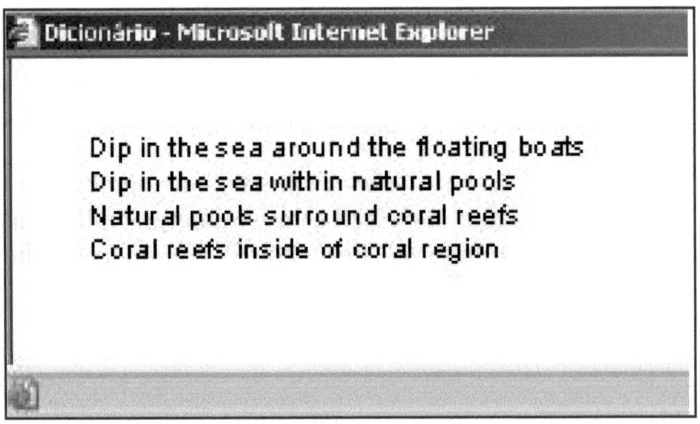

Fig. 6. Pop–Up window that shows detailed information about geographic features used in the Tourist Query.

b) A Biologist Query Submission

In this scenario we suppose that a biologist wants to know where he can find no consolidate substrates on the coral reef region. This information is not stored on the database like this.

In this example, the activate ontologies are O_{geo} and O_{bio}. So, we need to start to look for the "no–consolidate substrate" class in the O_{bio} ontology. As soon as this class is found on the biologist ontology, the system has look for class similarities on the definition of this class. The class "no–consolidate substrate" in the biologist ontology has no similar class in any other ontology. In other words, we can say that the "no–consolidate substrate" class is not related with any other class using the owl tag sameAs. So, due to the fact that the conceptual similarity was not found in the first level of the biologist graph, the comparative module will go down one more level, looking for if the "no–consolidate substrate" class has some relationships with others classes or if it has sub–classes.

The comparative module finally finds that the "no–consolidate substrate" class has sub–classes (relationship ISA) with the classes: "sand", "gravel" and "mud". So, we can say, for example, that sand is a no–consolidate substrate.

Now we have to search about similar classes with "sand", "gravel" and "mud" on the geologist ontology. All these classes are defined on the geologist ontology too. So, with the conceptual similarity found, the next step is to generate the query clause to be submitted to the geographic database. The Query Generator Module will construct this following SQL query clause:

```
select geom_areia AS geom_areia,
       geom_lama AS geom_lama,
       geom_cascalho AS geom_cascalho
from areia, cascalho, lama
AS foo USING UNIQUE oid USING SRID = -1
```

This query results will show on the map the localization of all sand, gravel and mud within the coral reef region, like we can see on Figure 7. On Figure 8 we present the detailed information about the geographic features involved in the query.

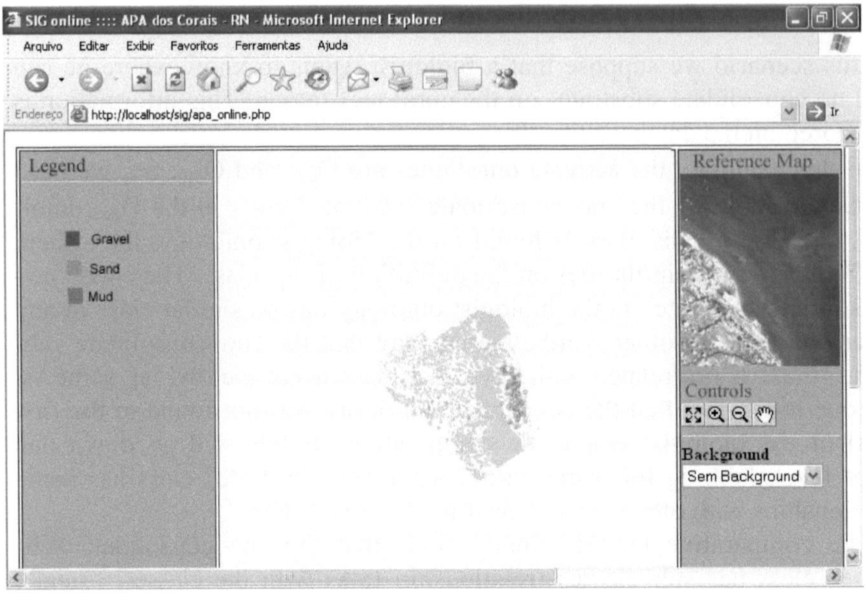

Fig. 7. Biologist Query Results about "no–consolidate substrate"

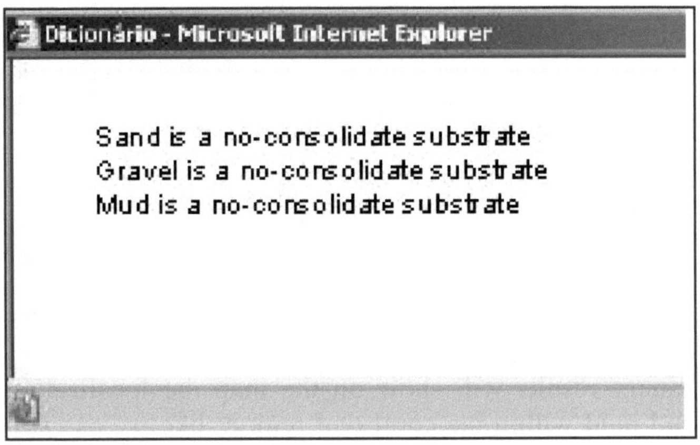

Fig. 8. Pop–Up window to show detailed information about the geographic features.

These two complex query clauses generated by the system are some examples of how this system could help different type of users to search geographic databases. We have some updates to do in this system, but it is important so say that it is already implemented and that it works [17].

5 Conclusion

This work presented an ontology–based mechanism to access a geographic database. This mechanism allows that different users' communities, through geographic ontologies, access the same database, without know his internal structure.

It was developed a semantic layer that integrates the geographic information, through the use of ontologies and through the definitions about the semantic similarity between classes, which have the same concepts, but different nomenclatures.

In this work we developed ontologies for the coral reef domain, based on three points of view of different communities: the geologists community, the biologists and the tourists [17]. Each class of each ontology has a similarity with some class of other ontology, but the terms used to define them could be different. The proposal ontologies can be used as a navigation and query tool for the users, supplying the semantics information desired.

For future work we could define a ranking of similarity between the classes of the ontologies. This ranking could help the synonym classes mapping and decrease the time for search the geographic database, and also avoid the use of terms that could not be of interest of the specific activated application.

It is important to point out that although the domain chosen is a little restricted, the proposed architecture can be adapted to any spatial domain of multidisciplinary interest. And, in this case, the return of these new applications with the use of ontologies, using this architecture presented in this proposal, could be very satisfactory [17].

References

[1] Breitman, K. K, Casanova, M. A. (2005) Desenvolvimento de Ontologias para Engenharia de Software e Banco de Dados: Um tutorial Prático. Tutorial presented at XIX Brazilian Symposiums on Software Engineering. Uberlândia – MG–Brazil.

[2] Cardoso, R. C., Souza, F. F., Salgado, A. C. (2005) Using Ontologies to Prospect Offers on the Web. In: 7th International Conference on Enterprise Infor-

mation Systems, 2005, Miami. ICEIS 2005 - Software Agent and Internet Computing. Lisboa - Portugal : INSTICC, 2005. v. 4. p. 200-207.
[3] Dongilli P., Franconi E., Tessaris S. (2004). Semantics driven support for query formulation. In Proceeding of the 2004 international workshop on description logics (dl'04), 2004.
[4] Egenhofer, M., Fonseca, F., Davis, C., Borges, K. (2000) Ontologies and Knowledge Sharing in Urban Gis. CEUS - Computer, Environment and Urban Systems. 24(3): 232-251
[5] Egenhofer, M., Herring, J. (1991). Categorizing Binary Topological Relations Between Regions, Lines, and Points in Geographic Databases. Technical Report, University of Maine, Orono.
[6] Fonseca, F. (2001). Ontology-Driven Geographic Information Systems. Phd Thesis, University of Maine.
[7] Franconi E., Tessaris S (2005). A unified logical framework for rules (and queries) with ontologies. In Rule Languages for Interoperability, 2005.
[8] Guarino, N. (1998) Formal Ontology and Information Systems. In: Formal Ontology and Information Systems (FOI'S 98). Italy, 1998.
[9] JENA – A Semantic Web Framework for Java - Jena 2 Ontology API. (2004). http://jena.sourceforge.net/ontology.
[10] Noy, N. F., Musen, M. A.: SMART: Automated Support for Ontology Merging and Align-ment. Workshop on Knowledge Acquisition, Modeling, and Management, Banff, Alberta, Canada, 1999.
[11] Open GIS Consortium, Simple Features Specification for SQL. Revision 1.1. May, 1999. http://www.opengis.org/.
[12] OWL WEB ONTOLOGY LANGUAGE GUIDE (2004). http://www.w3.org/TR/owl-guide/.
[13] PostGis, 2006. PostGis Geographic Objects for PostgreSQL. Disponível em: http://www.postgis.org. Accessed: 07/08/2006
[14] PROTÉGÉ ONTOLOGY EDITOR (2005). <http://protege.stanford.edu/>.
[15] Sondheim, M., Gardels, K., Buehler, K. (1999) GIS Interoperability. In: P. Longley, M. Goodchild , D. Maguire and D. Rhind. Geographical Information Systems - Principles and Technical Issues. John Wiley & Sons, New York, pp. 347-358.
[16] Surveys and Resource Mapping Branch – Ministry of Environment, Lands and Parks, British Columbia – Canadá (1994). Spatial Archive and Interchange Format: Formal Definition 3.1.
[17] Viegas, R. F.(2006) GeOntoQuery – Um Mecanismo de Busca em Bancos de Dados Geográficos Baseado em Ontologias. Master Thesis, Programa de Pós–Graduação em Sistemas e Computação.

A Method for Defining Semantic Similarities between GML Schemas

Angelo Augusto Frozza[1], Ronaldo dos Santos Mello[2]

[1]Departamento de Ciências Exatas e Tecnológicas
Universidade do Planalto Catarinense
[2]Departamento de Informática e Estatística
Universidade Federal de Santa Catarina

1 Introduction

The complexity and diversity of geographic data models found on Geographic Information Systems (GIS) allow for different ways to represent the same geographic reality. The increasing use of GIS on organizations has raised the need for georeferenced information interchange among autonomous and heterogeneous sources [9].

Information exchange among heterogeneous GIS presents incompatibilities both on syntactic and semantic levels. The syntactic level refers to the schema used on each system for data storage and documentation. Conflict resolution on this level is based on direct syntactic conversion of export and import formats. However, simple data transfer and re-formatting from a certain system onto another one cannot ensure that data may have any meaning for a new user. Interoperability among GIS requires semantic interpretation in order to explain concept correspondences among different systems [2].

Existing solutions for semantic interoperability among GIS usually apply standards such as Geography Markup Language (GML) [6]. The objective of GML is to offer a set of basic concepts, including a model of geographic features and a collection of geographic object metaclasses, that allows the user to structure and describe his/her georeferenced data.

In spite of the advances of the current GML version (3.1.1), semantic representation is still limited. The association of some ontological description with GML schemas may be a solution for this problem. In this context, OWL (Web Ontology Language) is the most recent W3C (World Wide Web Consortium) specification for ontology representation, being compatible with general Web architecture and particularly with Web Semantics [7].

This paper introduces a method for semi-automatic determination of semantic similarity between distinct GML schemas, using ontology as a basis for common knowledge. The contribution of this method is its support for the development of systems that exchange information among geographic databases with data semantics consideration. The domain of urban registration was used as a case study because it is an important GIS application domain that is few explored by related work.

Related work [1, 5] about GIS semantic interoperability focus on strongly tight environments, with emphasis on query translation. Instead of them, we work on geographic data integration in a specific application domain, considering that the data sources are not interconnected, but may perform frequently geographic data exchange in order to update common data.

This paper is organized as follows. Section 2 introduces our method. Sections 3 to 5 details its three parts: preprocessing, similarity determination, and mapping storing, respectively. Section 6 presents final considerations.

2 A Method to Determine Semantic Similarity

The proposed method (Figure 1) finds out semantic similarities between two distinct GML schemas: one representing data from a main GIS (GML_M), and other one representing imported data from a second GIS (GML_I). The geographic concepts considered by the method are described in a domain ontology, which is also useful in the definition of similarity scores. We do not support ontology update based on new concepts at the time of this paper. This talk will be treated by future work.

In the first step, we try to map GML_I elements to GML_M elements. It receives two inputs: the domain ontology and GML_I. GML_M is not considered because we assume that the method is applied in the context of the main GIS environment, and the semantic equivalences between the ontology concepts and the GML_M schema concepts had already been defined. With this assumption, we provide an incremental way of determining semantic equivalences among GML data. Elements from a new GML schema will always be compared against ontology concepts and the defined correspondences will, in turn, associate them to the elements of other GML schemas previously matched to ontology concepts.

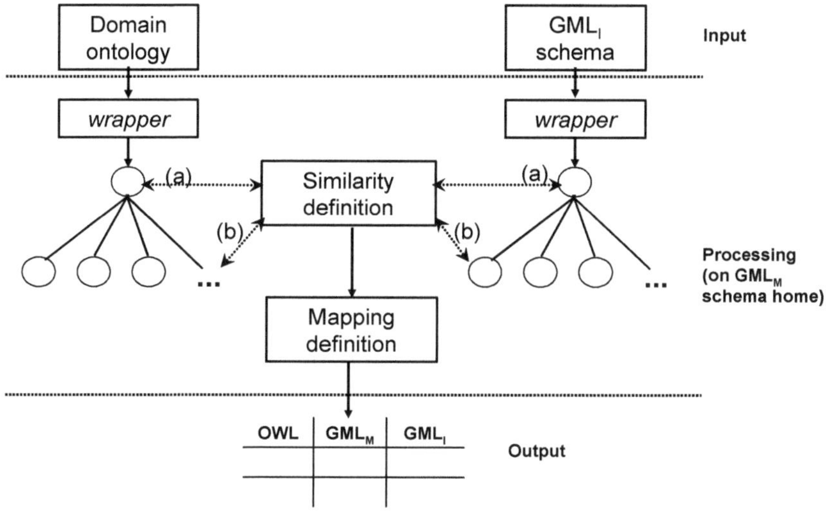

Fig. 1. Method overview

Two elements from distinct GML schemas are considered semantically similar when they hold a certain similarity score. This score is determined by metrics that consider the similarity among the element identifiers, as well as their attributes and relationships. The considered similarity metrics were adapted from the ones found in Dorneles *et al.* [4]:

- *Metrics for Atomic Values* (MAV): applied to simple data (see "b" simple element on Figure 1), such as strings and numbers. They depend on the application domain, i.e., they take the application data characteristics into account;
- *Metrics for Complex Values* (MCV): applied to the data structure (see "a" complex element on Figure 1). They may be distinctly applied to value sets (tuples) or collections.

These metrics was chosen because they define an appropriate taxonomy for XML data processing. In this case, an XML element is processed as a tree, considering that it may be an atomic or a complex element. Atomic elements contain simple data. Complex elements correspond to structures composed by other atomic or complex elements.

Table 1 shows the correspondences among OWL and GML concepts that are considered by the method for comparison purposes. The related concepts have the same intention on both languages.

Table 1. Mapping among OWL and GML concepts

OWL	GML
Class	Element
Property	Element (simple or complex) and attribute
Association	Hierarchy relationship
Specialization	Types derivation

Once semantic similarity between ontology and a GML_I schema are determined, the second step returns a mapping table that catalogs the found equivalences from the GML_I schema and the GML_M schema elements.

3 Input Data Preprocessing

The ontology and the GML_I schema must be translated into a canonical format in order to reduce the complexity of the similarity determination task. As both of them are XML structures, a tree data structure is used as a representation canonical format. The *wrappers* in Figure 1 are responsible to the generation of the canonical trees, which are exemplified in Figure 2.

Data in the ontology are organized into:

- class definition – denotes the name of the class;
- class properties –represents simple (strings, numbers, etc.) or complex (formed by other attributes) attributes and relationships. Moreover, attributes and relationships may receive one or more values;
- object instances – defines a synonym dictionary (see Section 4).

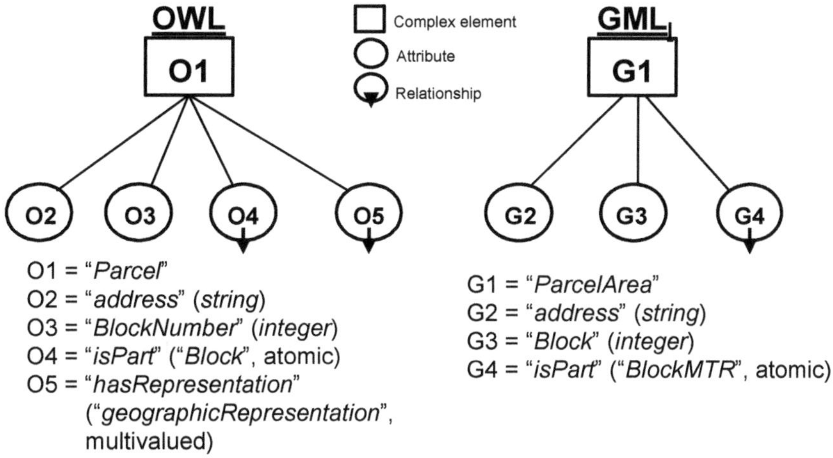

Fig. 2. Example of a canonical tree for an OWL element and a GML element.

On the canonical tree, the ontology description is organized hierarchically, i.e., properties from a C_i class (attributes and relationships) are structured as C_i child nodes. A canonical tree is generated for the GML_I schema in a similar way. In this case, each element becomes a non-leaf node on the tree, and leaf nodes represent its attributes and relationships.

4 Similarity Score Definition

Once canonical representations are generated, similarity scores between GML tree nodes and ontology tree nodes are defined. A GML node may have similarity with more than one ontology node. Thus, a similarity score is estimated through a numeric value from 0 to 1 in order to facilitate the definition of the best similar candidate.

Similarity definition is strongly based on a linguistic approach [8], i.e., it defines equivalence through text equality or similarity, yet considering element structures.

Our method considers a list of synonyms associated to each ontology class, with the format [SYNONYM, CLASS, LANGUAGE] (Table 2). The synonyms complement the ontology classes, identifying alternative known denominations for each term. This synonym list is obtained from instances of *Dictionary* class in the ontology, but it might be stored in a database. The synonym list was kept in the ontology in order to concentrate all knowledge about the application domain concepts in only one repository.

Table 2. Example of a synonym list

SYNONYM	CLASS	LANGUAGE
Lote	Parcel	Pt
Parcel	Parcel	En
Quadra	Block	Pt
Quarteirão	Block	Pt
Block	Block	En

The identifier of each GML element representative node (for instance, the name "*Block*") is first tested for equality against the synonym list. If one or more corresponding synonyms are found (for instance, "*Block*" = "*Block*"), a *structure similarity metric* (MCV) is applied to each positive result, in order to define the similarity score between nodes. Otherwise, a new search is done on the synonym list applying a *name similarity metric* (MAV) (for instance, "*CTMBlock*" = "*Block*"). If the similarity score is higher than an acceptable threshold defined by the user, the structure similarity metric is applied.

Several string similarity metrics are found in the literature [3], like *Jaro metric*, *Levenshtein distance*, and *Hamming distance*. In our method, it is possible to select one of these three metrics for defining similarities. Thus, we let the user free to choose the best metric to each situation.

The string similarity metric is applied to all simple data (MAV), which also consider data type equivalence between two attributes being compared. These metrics are detailed in the following.

Due to the nature of the handled data, the determination of structure similarity (complex elements) only requires the adaptation of a tuple similarity metric, as proposed by Dorneles *et al.* [4]. This is justified by the fact that the definition of a GML schema complex element is composed by its identification (name) and its properties (attributes and relationships), similar to a tuple in a relational table.

The metric definition is the following:

$$tupleSim(\varepsilon_p, \varepsilon_d) = \frac{\sum_{\varepsilon_p^i.\eta = \varepsilon_d^i.\eta} \left(sim(\varepsilon_p^i, \varepsilon_d^i) \right)}{\max(m, n)} \tag{4.1}$$

where:

- P: set of element nodes on the GML schema tree;
- D: set of class nodes on the ontology tree;
- ε_p: a P set node;
- ε_d: a D set node;

- n and m: number of children nodes from ε_p and ε_d, respectively.

The *tupleSim()* metric work as follow. Each ε_p^j node child of ε_p is compared to a ε_d^i node child of ε_d with same name ($\varepsilon_p^j.\eta = \varepsilon_d^i.\eta$) and same characteristics. Function *max()* returns the highest number of children between ε_p and ε_d.

We adapt this metric considering that an attribute can be either simple or complex, and handling relationships as complex attributes. In the case of complex attributes, the metric is applied in a bottom-up way, i.e., it starts on the last level of the GML element canonical tree onto the higher levels. Thus, a complex attribute on a higher level is handled as an atomic attribute, once its similarity score was defined on a previous iteration.

Besides, in order to compare a GML element child node to an ontology tree child node, two additional MAV metrics were defined in this work, given the type of the found node:

- *Metric for simple attributes*: compares attribute names (name similarity metric) and their data types (compatibility analysis):

$$attrSim(\varepsilon_p.\eta, \varepsilon_d.\eta) = \frac{nameSim(\varepsilon_p, \varepsilon_d) + typeSim(\varepsilon_p, \varepsilon_d)}{2} \quad (4.2)$$

  where:

  - ε_p: GML tree element child node;
  - ε_d: OWL tree element child node;
  - *nameSim*: attribute name similarity;
  - *typeSim*: data type similatiry.

- *Metric for relationships*: compares relationship names (name similarity metric) and cardinality similarity (1:1 – atomic; 1:n – multivalued):

$$relSim(\varepsilon_p.\eta, \varepsilon_d.\eta) = \frac{nameSim(\varepsilon_p, \varepsilon_d) + concSim(\varepsilon_p, \varepsilon_d) + cardSim(\varepsilon_p, \varepsilon_d)}{3} \quad (4.3)$$

  where:

  - *nameSim*: relationship name similarity;
  - *concSim*: concept similarity;
  - *cardSim*: cardinality similarity.

We exemplify the application of *tupleSim()*, *attrSim()* and *relSim()* metrics on the elements *Parcel* and *ParcelArea* in Figure 2. First of all, similarities between atomic attributes are calculated (only the correct combinations are presented):

- *sim2* = *attrSim*(G2, O2) = 1
- *sim3* = *attrSim*(G3, O3) = 0,95

Then, similarities between relationships are obtained:

- *sim4* = *relSim*(G4, O4) = 0,98

With these results, the final similarity score between both elements is calculated:

- *sim1* = *tupleSim*() = (1 + 0,95 + 0,98) / 4 = 0,73

Because our method is semi-automatic, further user validation is considered in order to accept or reject the obtained scores.

5 Mapping Catalog

The last step of the method is responsible to catalog the found correspondences (mappings) between OWL and GML elements. These mappings are stored in two table sets in a relational database schema:

- The first table set keeps information related to the GML_I schemas (metadata), such as schema supplier identification, URL, responsible, schema version, languages, among others;
- The second table set keeps the correspondences. It is considered that there may be a similar concept on the ontology for each element on the GML_M schema. Thus, GML_I schema elements and similarities are related to concepts from the GML_M schema and from the ontology, using two tables. Figure 3 presents an example for such a catalog.

OWL x GML_M

ID	Ontology Concepts	GML_M	GML_M similarity
1	Block	Block	1
2	Parcel	ParcelDesigned	0,92

GML_M x GML_I

GML_M	GML_I	GML_I similarity
1	BlockMTR	0,94
1	BlockDesigned	0,90
2	ParcelMTR	0.96

Fig. 3. Mapping catalog

6 Conclusion

Considering that geographic data exchange occurs mainly among domains with some semantic affinity, geographic data are better defined semantically on a specific domain than through domain generalizations.

This paper proposes a solution for the problem of semantic interoperability among GML schemas in the context of urban registration. This solution considers equivalence determination using knowledge represented through a domain ontology. The method is semi-automatic because it does not discard user intervention for validation purposes or for cases when a GML element is associated to more than one concept in the ontology.

Related work focus on translating queries executed on closely interconnected heterogeneous environments. Our method considers another scenario: small municipalities, whose geographic data are available on many institutions, such as city halls and sanitation companies. In this scenario, we have several institutions that do not dispose of technical and financial support to make their data available. However, if working as a group, they are able to promote data interchange through a mechanism for identifying similarities among them. Thus, it is possible to socialize urban geographic data and develop new services for the community in general.

Future work include the application of the method for other domains and the specification of a complete system for mediation-based queries by similarity, as well as the integration of data instances coming from heterogeneous geographic data sources.

References

1. Brauner DF, Casanova MA, Lucena CJP (2004) Geo-Object Catalogs to enable Geographic Databases Interoperability. In VI Brazilian Symposium on Geoinformatics Proceedings, pp 235-246. INPE.
2. Câmara G, Thomé R, Freitas UM, Monteiro AMV (1999) Interoperability In Practice: Problems in Semantic Conversion from Current Technology to OpenGIS. In INTEROP 1999 Proceedings, pp 129-138.
3. Chapman J (2006) Sam's Strings Metrics. <http://www.dcs.shef.ac.uk/~sam/stringmetrics.html>.
4. Dorneles CF, Heuser CA, Lima AEN, Silva AS, Moura ES (2004) Measuring Similarity between Collection of Values. In VI International Workshop on Web Information and Data Management Proceedings, pp 56-63. ACM.
5. Morocho V, Pérez-Vidal L, Saltor F (2003) Semantic Integration on Spatial Databases: SIT-SD prototype. In VIII Jornadas de Ingeniería del Software y Bases de Datos Proceedings, pp 603–612.
6. OGC (2003) Geography Markup Language (GML) Implementation Specification 3.0. Open GIS Consortium.
7. OWL (2006) Web Ontology Language <http://www.w3.org/2004/OWL/>
8. Rahm E, Bernstein PA (2001) A survey of approaches to automatic schema matching. The VLDB Journal, 10(4):334-350.
9. Zhang J, Guan J, Zhang J, Chen J (2004) Geographic Information Integration and Publishing Based on GML and SVG. In IV International Conference on Computer and Information Technology Proceedings, pp 764-769. IEEE.

Interoperability among Heterogeneous Geographic Objects

Victor H. M. Azevedo[1], Margareth S. P. Meirelles[1,2],
Rodrigo P. D. Ferraz[2], Antônio Ramalho-Filho[2]

[1] Departamento de Engenharia de Sistemas e Computação
Universidade do Estado do Rio de Janeiro
[2] Centro Nacional de Pesquisa de Solos
Empresa Brasileira de Pesquisa Agropecuária

1 Introduction

The integration of geographic objects stored as distinct data sources with heterogeneous syntactic and semantic structures has been target for researchers that with computing systems in distributed environment of geoprocessing over the last years. This fact happens owing to an increasing need for information exchange processed by those generators of geographic data.

Many initiatives have been performed in order to attain the interoperability between institutions intending to exchange information among them. According to Fonseca and Egenhofer [1], the first attempts to obtain interoperability on Geographic Information Systems (GIS) have been done through direct exchange between software makers. At present, specialized technicians and institutions working with Geomatics form the OpenGeospatial Consortium (OGC), aiming to define a set of standardized specifications for interoperability in GIS. The GeoBR initiative [2], proposes an uniform schema, by using pre-defined elements and that includes metadata, projections, geometry and attributes that can be accessed through just one programming interface.

The cooperatives on geographic data described by Câmara et. al. [3] are an emergent solution for managing a great deal of information, so as to allow its cooperative use by government or private agency.

According to Hartman [4], in projects which involve a heterogeneous environment, the higher costs refer to data acquisition. It represents 60 – 80% of total the cost of GIS implementation. Based on this assertive the high cost for collecting and producing geographic data is an incentive factor to the interoperability of spatial information already produced by several institutions. Upon this scenario, the automation of the interoperability between geographic objects within distributed environment becomes a powerful tool, which the main objective is to make feasible the cooperation among the institutions that produce spatial information.

To attaining the complete geographic interoperability of objects on a way that they can be interpreted under the same optic, is not an easy task [2]. According to Casanova [5], the complete integration either of data structure arranged or the significance of interpretation should resolve the incompatibility under three levels: format and structure, syntactic, and semantic.

Much effort has been spent to provide computing mechanisms for standardizing formats and giving semantic and syntactic uniformity to the geographic objects. According to Lima et al [2], nowadays it is unquestionable the use of Extensible Markup Language (XML) (http://www.w3.org/xml) as pattern to data exchange. The OGC provides a set of specifications for standardizing interoperability process between different data formats based on the technology XML. The Geographic Markup Language (GML) (http://www.opengeospatial.org/standard/gml) can be considered the main of them, owing to its use in many other specifications. It was conceived with the objective of representing geographic information including either spatial information as non spatial ones. According to Davis Jr (2005), the objective of GML is to offer a set of rules in which the users can define their own language in order to describe geographic objects they intend to handle.

The OGC provides an architectural framework that defines, through formal specifications, the scope, objectives and behavior of a series of web facilities called OpenGIS Services Framework. The objective of such a framework is to provide a mechanism that is able to guarantee the interoperability among institutions by using the internet.

The OGC specifications do not include any concern within semantic aspects related to the interoperability in terms of its approach; neither follow the recommendations from the Consortium W3C when using semantic web, as it is said by Davis Jr. et. al. [6].

The term 'ontology' has nowadays been used in information sciences in order to represent a "formal explicit specification of a shared conception" [7].

The use of ontologies as a strategy for representing knowledge about a given interest domain provides semantic schema that has shown efficacy, once it is possible to specify on an explicit and formal way the terms of a domain as well as the relationship among them. The OWL (http://www3.org/TR/owl-ref) is a language that has the objective of defining, publishing and sharing ontologies on the web and can make feasible the semantic interoperability.

Considering the scenario above shown, this article proposes a methodological approach able to automate the interoperability among available geographic objects at several institutions, aiming to reduce the costs of this integration and making more agile the process of decision making.

This methodology is been applied to the Agroecological Zoning Project for Palm oil in deforested areas in the Brazilian Amazon for biofuel production. The carrying out of this project requires soil information taken from multiple integrated sources

2 Methodology

The syntactic, semantic heterogeneity and spatial data structure must be taken into account when performing such integration. To solve this problem Klien et al [7], proposes a kind of architecture based on geographic web services that uses the BUSTER System (http://www.informatik.unibremen.de/agki/www/buster/new) to integrate information on storms in forests. The BUSTER system is based on ontologies used for web search and information integration within heterogeneous distributed environment [7]. Also, alternatively, this methodology proposes the use of specification from GML and OGC web services as a technological mechanism to the syntactic integration as well as structure among heterogeneous geographic objects. Besides that, the syntactic ontologies and domain description in OWL language format, can define syntactic semantic aspects providing a technological framework of the knowledge representation about that domain.

The methodology used has six phases (Fig. 1):

- Definition of OGC Services Oriented Architecture (GSOA), with definition of Geographic Objects Servers (GOS) and the Integration Servers (ISGeo);

- The use of Web Feature Service (WFS) to provide the geographic objects from each data source through GOS;
- Construction of the knowledge bases with the creation, under an OWL language, of the domain ontology ones intend to exchange information with, through knowledge engineering process;
- Publication of the services from each GOS within the ISGeo;
- Disposal of the integrated information on a service of WFS and Web Map Service (WMS) format;

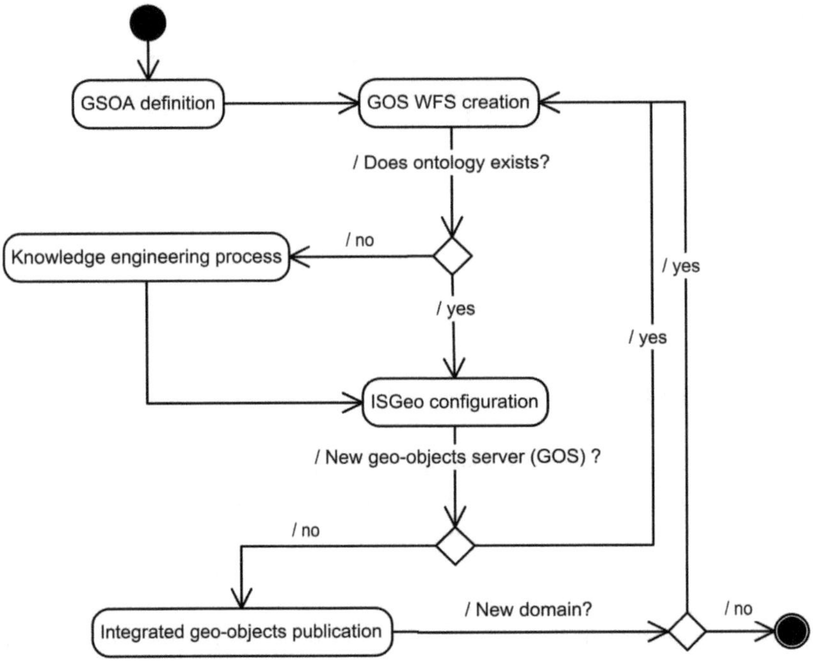

Fig. 1. *Activities diagram of the methodology for geo-object integration*

2.1 OGC Services Oriented Architecture (GSOA)

Two types of different servers must be implemented under the architecture being defined: Geographic Objects Servers (GOS) and Integration Servers (ISGeo)

The GOS are responsible for providing geographic information from each institution involved in the integration, transforming data into a GML pattern format. These data on GML format must be provided through WFS services. The ISGeo, in turn, are responsible for registering the WFS services that are available in GOS and for unifying the syntax and semantic of each element of these services in such a way that all of them can be interpreted identically. For doing so, the ISGeo must allow to register the ontology of domain and provide a mechanism for correlating the available elements in GOS with this related ontology.

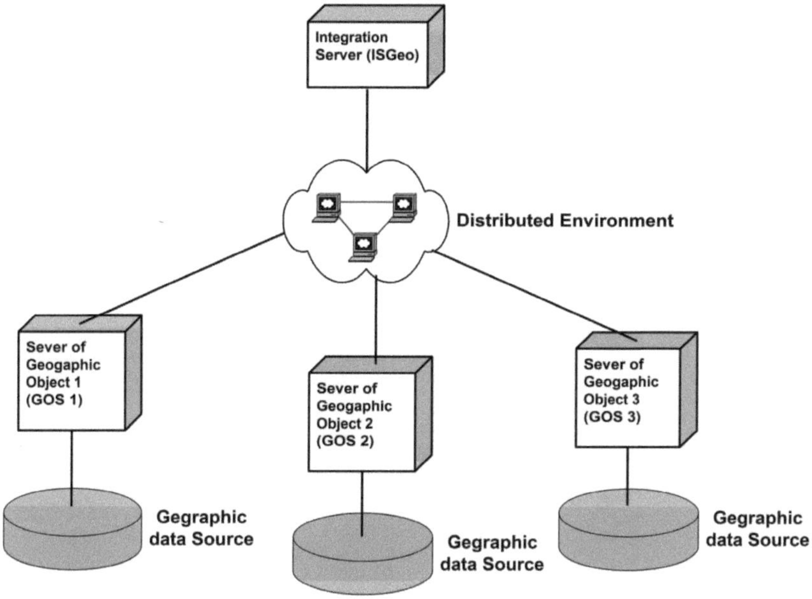

Fig. 2. *Example of OGC Services Oriented Architecture (GSOA)*

2.2 Creating WFS Services

After the architecture is defined and the servers of geographic objects are known, it is necessary to make automatic the process of search and publication of this stored information in each institution. For doing that, in each servers of geographic objects, are defined the WFS needed to turn the available geographic objects into GML format.

Each different form of representing spatial information used by institution is, initially, converted into GML and, afterwards disposed to whom may ask for the service in this format. Some SIGs and map servers already have compatible tools with WFS specifications belonging to the OGC consortium and they convert the geographic objects to the respective representation in GML.

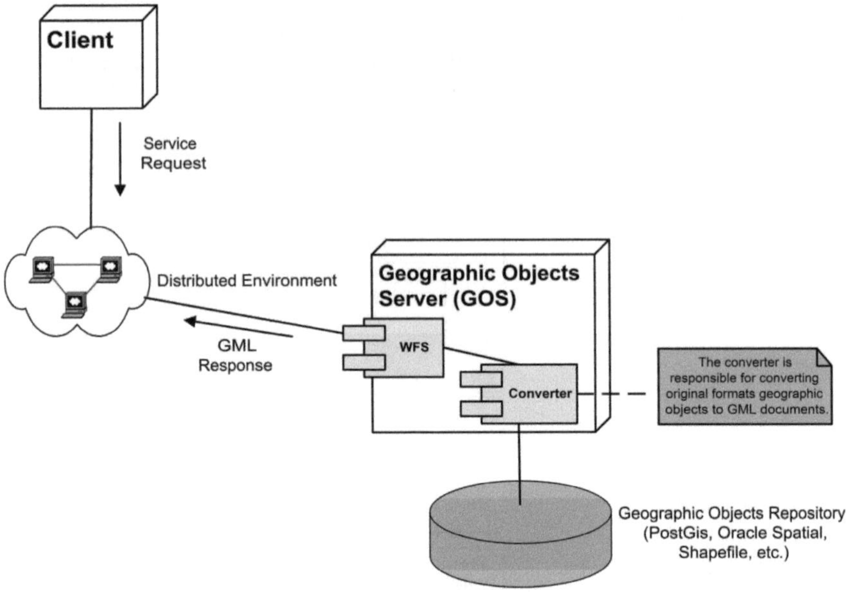

Fig. 3. *The functioning of WFS Services*

2.3 Knowledge Engineering Process

This phase proposes performance of a knowledge engineering process, which objective is to generate ontologies to represent the domain terms and its relations on formal basis. It is provided, in this way, a conceptual schema of reference that makes possible the available geographic information syntax and semantic uniformization.

Two stakeholders are identified to perform the task to represent the knowledge about the domain. In this methodology are entitled: Knowledge Engineer and Domain Expert. The Domain Expert is a professional that withhold the knowledge and has practical experience on dealing with the area to which he intend to model. The Knowledge Engineer is the professional who has as mission to model the knowledge on domain as from the acquired information from the specialist and formally represents him through the creation of an ontology. Interview, bibliographic review, workshops with specialized groups and other techniques must be used in this phase in order to give support to the understanding process as well as in the domain modeling. The Knowledge Engineer and Domain Expert must interact themselves with the objective of formalizing a standardized language to represent the terms of domain.

Another important aspect during the development of a knowledge base is the relationship among ontologies aiming at the reuse. On beginning the process, one must realize that the ontology which is being created can use either other ontologies which have already been modeled or reused late on by other ontologies. At present, there already exist available ontologies that can be imported and used as edition tools of ontologies, and the trend is that there is an increasing over the production and publication of ontologies as time goes by, stimulated mainly due to the advance on research related to web semantic.

The knowledge base must be compounded by ontologies necessary to form a syntactic and semantic referential about the area of interest. The domain must be represented in Ontology Web Language (OWL), being desirable the generation of the entire semantic model in web page. The OWL will allow the manipulation of domain terms through computing programs aiming at the correlating terms, while the publication of ontology within web pages make easier the sharing of information among users.

2.4 Publication of Services on the Integration Server

The objective of this phase is to create an automated mechanism for unifying available geographic data from various data sources through the server of integration. Using the registered information the SI might provide the results of the integration within GML format, following the ontology, that is, the reference model defined by the knowledge base.

The integration server must offer the following functionalities:

- Register of domains that form the knowledge base with its respective ontology in OWL;
- Publication of the WFS Services with the geographic objects that should be integrated, by using a pre-determined metadata structure;
- Syntactic and semantic correlation among conceptual schemas used in WFS services and a reference schema defined by the ontology in OWL.

By this way, the server configuration is performed through the location register of the WFS services available in each data source network, together with the information on the services on a metadata basis as well as the syntactic and semantic correlation among conceptual schemas. With this configuration already built, the server of integration might, in fact, unify the available geographic objects in each data source registered on it.

2.5. Making available integrated geographic objects

The unified geographic objects must be disposal in GML format through WFS services. Besides the GML format publication, a visual representation of the unified geographic data must be available through WMS service. WFS service will allow the unified geographic objects to work as a new data source, while WMS will allow the analyses and the manipulation of the map generated by the unified geographic objects.

3 Case Study: Heterogeneous Soil Database Integration

The methodology has been applied in soil domains and land suitability aiming at evaluating the land potential for palm oil to produce biofuel in deforested areas in the Brazilian Amazon, under the Agro ecological Zoning Project for Palm Oil, coordinated by Embrapa Soils. Thus, it was used information on soil obtained from SIPAM, National Center for Soil Research (Embrapa), and The Brazilian Institute for Geography and Statistics (IBGE).

Land evaluation for a given crop or produce is done based on the interpretation of data from natural resources and matched with crop requirements. According to this assertive, there is a real need to integrate different available data source. In the context of this document, evaluating land, considering all its criteria and procedures, refers to the 'Brazilian System for Evaluating Land'[8], which is the method recommended by Embrapa for soil surveys interpretation.

Two data sources presenting heterogeneity in their schemas (Table 1) were chosen so as to exemplify the syntactic semantic structural integration of soil geographic information.

A single architecture was defined, with two geographic object servers, The GOS-SIPAM and the GOS-IBGE, as well as a server of integration, the SIGeo. Afterwards, for each GOS, it was created the WFS service through the GEOSERVER tool (http://docs.codehaus.org/display/GEOS).

The knowledge base on soil agricultural suitability and soil attribution (Fig.4) was developed so as to establish a semantic reference for integration. Using data from each source, available in GML format, and tables which correlate the heterogeneous source schema terms with the similar semantic terms from the reference ontology, the data were integrated generating a single map with the geographic objects from both institutions. The map was generated with a structure based on the ontology used as semantic reference.

Table 1. *Example of heterogeneity found in the soil databases*

Heterogeneity	SIPAM	IBGE
Format	Postgis Spatial Database	ESRI Shapefile
Syntactic	Geo	SHAPE
Semantic	Relief mapping unit polygon in spatial table	Relief mapping unit polygon in shapefile
Syntactic	cd_letra_simb	SIMB_UNID
Semantic	Mapping unit symbol	Mapping unit symbol
Syntactic	cd_textura_1h1	GRUP_TEXT
Semantic	Superficial horizon texture	Texture group
Syntactic	cd_relevo_1	FS_RELEVO
Semantic	Predominant relief	Relief phase
Syntactic	cd_pre_casc_1_t1h1	FS_PEDREG
Semantic	Presence of gravel in the superficial horizon	Stoniness phase

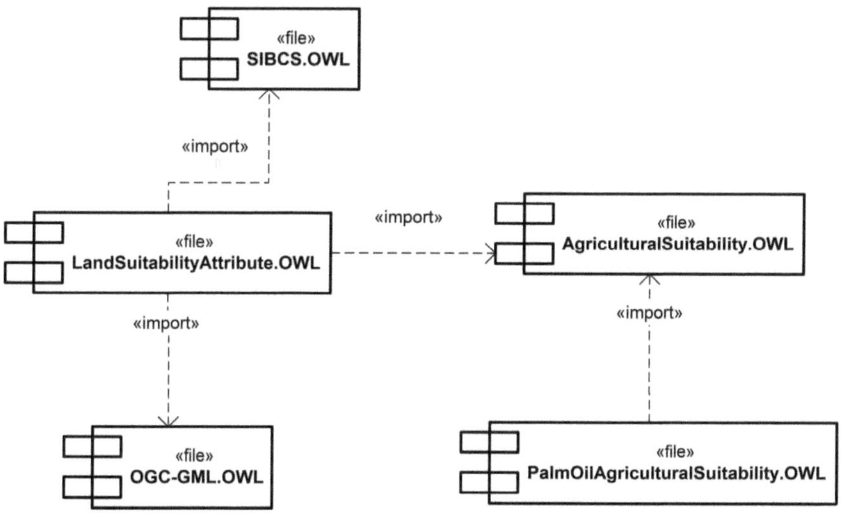

Fig. 4. *Knowledge Base*

4 Conclusions

With a heterogeneous simulation environment and the utilization of schemas and real conceptual information, the objective of format unification was reached through the use of WFS services. The GML format, based on

XML, was adequate for heterogeneous geographic information changing. The use of a service publication tool WFS, along with the automatic conversion of the data source original formats to GML, makes the geographic object conversion task to a standardized format much simpler.

The use of the base knowledge as well as the OWL ontology served as a reference schema for the semantic and syntactic integration among objects. This base knowledge was also used to validate the proposed methodology and it can also subsidize other agricultural field applications.

References

[1]Fonseca, F. T. and Engenhofer, M. J. (1999) Sistemas de Informação Geográficos Baseados em Ontologias, National Center for Geographic Information and Analysis, Department of Spatial Information Science and Engineering, University of Maine, Orondo, ME 04469-5711, USA.

[2]Lima, P. ; Câmara, G. and Queiroz, G. (2002) GeoBR: Intercâmbio Sintático e Semântico de Dados Espaciais", INPE, São José dos Campos, SP, Brasil.

[3]Câmara, G.; Casanova, M. A.; Hemerly, A. S.; Magalhães, G. C. e Medeiros, C. M. B. (1996) Centros de Dados Geográficos, Anatomia de Sistemas de Informações Geográficas. Rio de Janeiro, Brazil. pp. 167- 173.

[4]Hartman, R. (1998) GIS Data Conversion – Strategies, Techniques and Management, Onword Press, New York, USA.

[5]Casanova, M. A.et. al. (2005) Integração e interoperabilidade entre fontes de dados geográficos, Bancos de Dados Geográficos, Editora MundoGeo, Rio de Janeiro, Brazil, pp. 305- 340.

[6]Davis Jr., C. A. et al. (2005) O Open Geospatial Consortium, Bancos de Dados Geográficos, Casanova, M. et al., MundoGeo. Rio de Janeiro, Brazil, pp. 367-383.

[7]Klien, E., Lutz, M , Kuhn, W. (2004) Ontology-Based Discovery of Geographic Information Services – An Application in Disaster Management, 7th Conference on Geographic Information Science (AGILE 2004).

[8]Ramalho-Filho, A. and Beek, K. J. (1995) Sistema de avaliação da aptidão agrícola das terras. 3rd. edn. rev., EMBRAPA-CNPS ,Rio de Janeiro, Brazil, 65p.

Web Service for Cooperation in Biodiversity Modeling

Karla Donato Fook[1,2], Antônio Miguel Vieira Monteiro[1], Gilberto Câmara[1]

[1] Divisão de Processamento de Imagens
Instituto Nacional de Pesquisas Espaciais
[2] Departamento Acadêmico de Informática
Centro Federal de Educação Tecnológica do Maranhão

1 Introduction

The biological resources of biodiversity support essential services and sectors, such as Food and Agriculture, Water, Pharmaceuticals, Medicine and Waste treatment [12]. Biodiversity information is fundamental for the scientific, education and government communities, for preservation of the world's fauna and flora, as well as in decision-making processes during the urban and regional planning. Scientists working with biodiversity information employ a wide variety of data sources, statistical analysis and modeling tools, and presentation or visualization software. These resources may be available on various local and remote platforms [24].

The collaboration among researchers involves interaction between scientific models and their implementations, programs aggregation and experiments results and the exchange of data [19]. Researchers working with the species distribution modeling, help the analysis and solution of problems, such as the forecast of species distribution, impact of climatic changes and problems related to expanding invader species. A computational infrastructure that allows sharing the knowledge contained in the modeling results can help scientists proceed in their studies, apply consolidated knowledge to solve new problems and obtain new knowledge.

The focus of this approach is to support a form of collaboration in species distribution modeling network, and support distributed applications on spatial data for this modeling. A useful infrastructure is that in which two or more users collaborate making available modeling results, independently of the application and platform. We propose to develop WBCMS (Web Biodiversity Collaborative Modeling Service), a web service that allows the creation, cataloguing and recovery of data and context of modeling results of species distribution, here called models instances. The service is in its early phase of development.

In this paper, we discuss the existent challenges for obtaining a computational infrastructure that supports cooperation among users of species distribution modeling networks. The remainder of this paper is structured as follows. Section 2 focuses on some of these challenges and approaches. Section 3 presents a WBCMS description and Section 4 describes an early experiment. Finally, section 6 concludes the paper and highlights some future research directions.

2 Challenges and Approaches of the Biodiversity Informatics and GI Web Services

Biodiversity data access, through new software tools, web services and architectures, brings opportunities and dimensions for new approaches in the ecological analysis, predictive modeling, and synthesis and visualization of biodiversity information. To integrate initiatives in an organized and global approach that build and manage resources of biodiversity information through collaborating efforts is a challenge in the biodiversity informatics [6, 16]. Biodiversity data access networks aim to make them available on the Internet. Some examples are listed below:

- GBIF[1] – Global Biodiversity Information Facility: Promotes development and adoption of standards and protocols for documenting and exchanging biodiversity data. [10, 18];
- SpeciesLink[2]: Distributed Information System that integrates primary data from scientific biological collections of São Paulo State, observation data of Biota/FAFESP3 Program and others [7];

[1] http://www.gbif.org/

[2] http://splink.cria.org.br/

[3] http://www.fapesp.br/

- Lifemapper[4]: Provides an up-to-date and comprehensive database of species maps and predictive models using available data on species' locations [22];
- MaNis[5] – Mammal Networked Information System: Development of an Integrated Network for Distributed Databases of Mammal Specimen Data;
- HerpNet[6] – Reptiles and Amphibians of Iowa and Minnesota: Collaborative effort by natural history museums to establish a global network of herpetological collections data;
- FishNet2[7] – Distributed Information System for Fish Networking: Distributed Information System to link the specimen records of museums and other institutions in an information-retrieval system;
- ORNIS[8] – ORNithological Information System: Expands on existing infrastructure developed for distributed mammal (MaNis), amphibian and reptile (HerpNet), and fish (FishNet2) databases.

GIS technology is moving from isolated, standalone, monolithic, proprietary systems working in a client-server architecture to smaller web-based applications [4, 9]. However, there are several challenges in this area. The architectures for workflow creation and managing; software and middleware development; user interfaces; protocols for data queries; analytical and modeling tools and Grid Networking applications [17]. Some approaches dealing with these subjects have been presented in the literature.

- Spatial Data Integration in Web [4, 15, 21]
- Chaining Static and Dynamic Web Services [1, 3, 5, 23]
- Collaboration and Grid Networking applications in GI Web Services [11, 13, 14, 19, 20, 25]

The cooperation and integration process of spatial data are linked. Pinto et al. (2003) [21] extended the architecture of data integration, which provides services to find, share and publish sources of data through the Web.

4 http://www.lifemapper.org/
5 http://manisnet.org/
6 http://www.herpnet.org/
7 http://www.fishnet2.net/index.html
8 http://ornisnet.org/

The collaborative Project SpeCS uses this architecture and has been used in decision-making processes [21].

Alameh (2001) [2] approaches chaining GI Services on the Web. She proposes architecture for the building of extensible infrastructure that supports the dynamic linkage of distributed services. This infrastructure facilitates the integration of GIS data providers with other information systems. Bernard et al. (2003) [5] propose the static linkage of GI Web Services to build up a more a complex task. The work was applied in estimating road blockage after storms. Tsou et al. (2002) [23] presented a dynamic architecture for distribution of Geographical Information Services with Grid Networking Peer-To-Peer technology. A framework based on existent languages, computational architectures and web services was implemented.

However, the mentioned approaches do not aim at making available the processing results, in this case a model, to the end-user community. Our approach attempts to fill in this gap by supporting the cooperation amongst modelers in a species distribution modeling network through the availability of the very own modeling outcomes. This approach involves an integrated view that brings together a workflow approach for chain processing, the definition of protocols for negotiating models and the handling of the spatial data.

3 WBCMS – Web Biodiversity Collaborative Modeling Service

This approach considers a distributed environment in which researchers perform the species distribution modeling in their station, and wish to cooperate with other users of a biodiversity network through the results of their modeling. WBCMS supports the cooperation in a modeling network through sharing modeling results of species distribution. This service helps researchers to find answers for issues such as:

- “Which are the modeled species?”
- “Where did the data come from?”
- “What are the used environmental variables?”
- “What is the used algorithm?”
- “How to visualize the model?”
- “If I have a problem, how can I look for similar results?”

WBCMS allows the creation, cataloguing and recovery of models instances with the use of a catalog. A model instance holds the conceptual information about the model and their generation, such as the input data, the modeling algorithms and its parameters. A model instance holds a metadata which is generated at run time together with the model results itself.

The researcher performs the modeling and calls the service to generate and catalog that particular model instance. Then, other researchers in the network can access the instance of that model. The Figure 1 shows the cataloguing and access use cases.

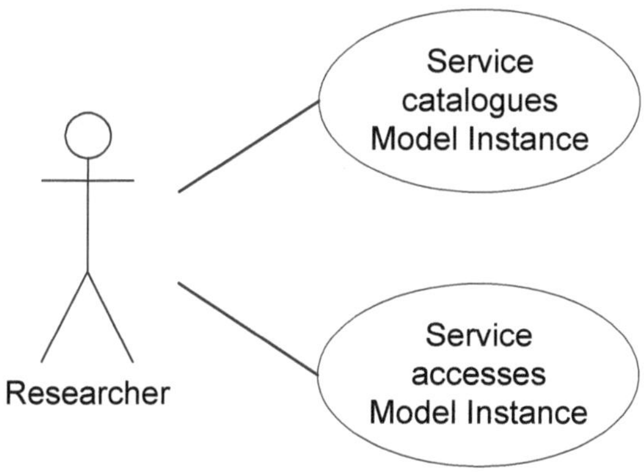

Fig. 1. Catalog and Access of Model Instance

After performing the modeling, the researcher invokes WBCMS to make available the results in a catalog. In a first moment the researcher calls the service to create and to catalog the model instance. For that, the service receives and prepares the modeling data, composes the model instance and inserts it in the catalog (Figure 2).

Figure 2 shows the use case of the cataloguing process. Primary actor of the use case is the Researcher, which launches the service. The "Success Warranty" indicates a successful trial. The extensions describe the handling of exceptions over the main scenario. In another moment, we considered that certain Researcher wants to consult model instances of a catalog for its studies. In that case, he calls the service to access the catalog. Such service refers to the way in which a researcher can access the model instances. Figure 3 displays this use case.

Use Case: Service catalog Model Instance

Primary Actor: Researcher

Scope: Species Distribution Modeling Network

Stakeholders and Interests:

- Researcher - wants to catalog the result of his/her modeling;

Success Warranty: the model instance was generated and saved into catalog

Trigger: Researcher calls the web service

Main Success Scenario:

1. The Researcher selects the web service to catalog the model instance
2. The Service prepares the environment to perform the modeling algorithm
3. The Service creates the structure with model's data and metadata to compose the model instance
4. The Service inserts the model instance generated in the catalog

Extensions:

2a. Specimens data (local and/or remote) or environmental variables are not available

2a1. The Service shows message and cancels the Service's request

Fig. 2. Use Case: Catalog Model Instance

Use Case: Researcher accesses model instance

Primary Actor: Researcher

Scope: Species Distribution Modeling Network

Stakeholders and Interests:

- Researcher – wants to access the model instance;

Success Warranty: the model instance was retrieved and visualized

Trigger: Researcher calls the web service

Main Success Scenario:

1. The Researcher selects the web service to recover the model instance
2. The Service recovers the model instance
3. The Researcher visualizes the model instance

Extensions:

2a. Model instance isn't cataloged

2a1. Service shows message and it restarts search process

Fig. 3. Use Case: Researcher accesses Model Instance

Figure 3 describes the use case for the access to model instances. The primary actor of this use case is also the Researcher, which calls the service when he wants to consult some model instance. The "Success Warranty" indicates when the service obtains success. The extensions describe what happens if the model instance isn't in the catalog. The following subsection describes the WBCMS architecture.

3.1 Architecture

Web services encapsulate the underlying applications and publish them as a service that can be remotely accessed. We propose a web service which consists of two components, and contemplates data and processes services. Each WBCMS component has a set of services to perform cataloguing and access to model instances. The cataloguing component, denominated CatalogModService, contains a set of services that deals with establishing the necessary data for composing a model instance and inserts it in the catalog. The access component is a set of services that perform the search, recovery and visualization of the existent model instances in the catalog. This component is denominated AccessModService. Figure 4 shows the WBCMS architecture.

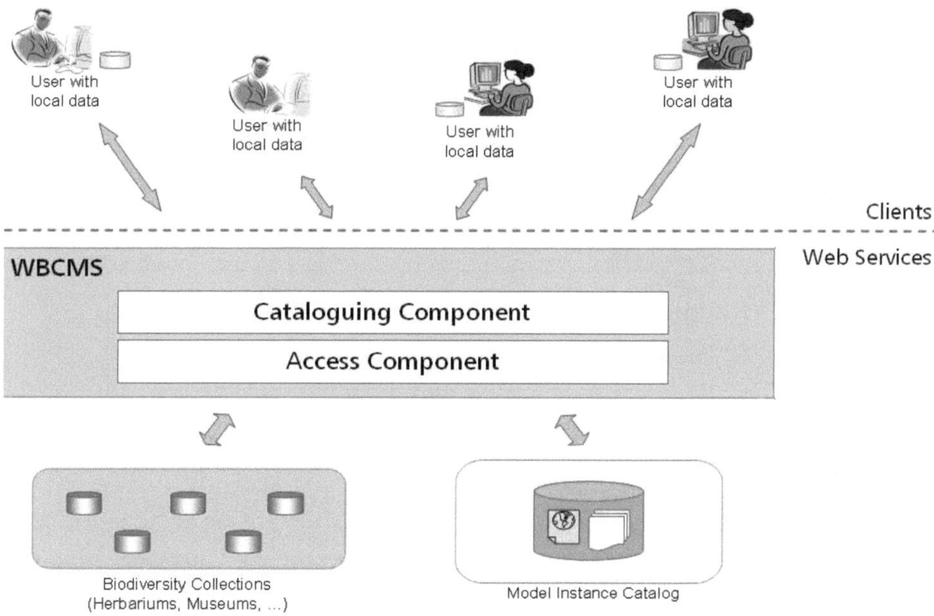

Fig. 4. WBCMS architecture

Figure 4 presents an architecture which considers that client uses local and remote data, performs the species distribution modeling in his/her PC, and can use the WBCMS in two different and independent moments. Figure 5 exhibits the two components.

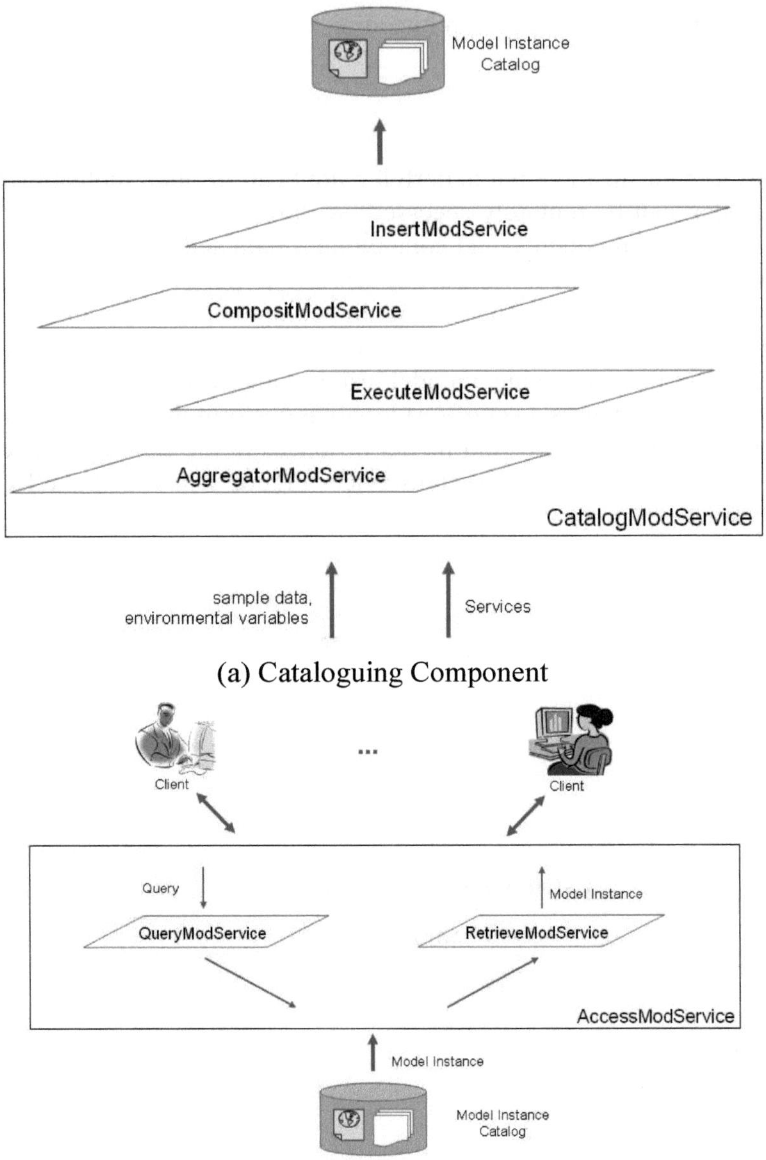

(a) Cataloguing Component

(b) Access Component

Fig. 5. WBCMS Components

Figure 5.a displays the services of the cataloguing component. The Services foreseen for that component are the AggregatorModService, which receives and prepares the modeling data; the ExecuteModService, which prepares data regarding the modeling algorithm; the CompositModService, that composes the model instance and creates the structure with the data and metadata of the resultant model and the InsertModService, which executes the cataloguing of the model instance. Figure 5.b shows the services of access component: the QueryModService, that performs the search of the model instance requested by the user, and the RetrieveModService that prepares the model instance, if it exists into catalog, and makes available it for the client.

The WBCMS is compliant with OGC[9] (OpenGis Consortium Inc.) and W3C[10] (World Wide Web Consortium). The service in development will be evaluated through a case study. The experiments will be applied inside the OpenModeller[11] Project context. OpenModeller is a spatial distribution modeling tool, which perform modeling and generate models through an algorithm that receives as input a set of occurrence points (latitude/longitude) and a set of environmental layer files. The outcome of the modeling process is projected into a geographic grid producing a georeferenced map representing the spatial distribution for that particular species over that particular geographic region [6, 8, 16].

4 Initial experiments

As an initial experiment, we built a prototype of the service to start an incremental process of development. We used Apache server, PHP and MySQL for developing the experiment.

The OpenModeller's desktop interface performs the species distribution modeling. It creates several files containing parameters, modeling data and information about the executed algorithm. These files have different formats such as .cfg, .html, .xml, .tif and .png, among others. In this experiment, the files are loaded in the server and some of them can be visualized by clients. A database contains the files' names that can be visualized, with information regarding the available algorithms in openModeller. However the model instance isn't generated. Figure 6 displays the general idea of the experiment.

[9] http://www.opengeospatial.org/
[10] http://www.w3.org/
[11] http://openmodeller.cria.org.br/

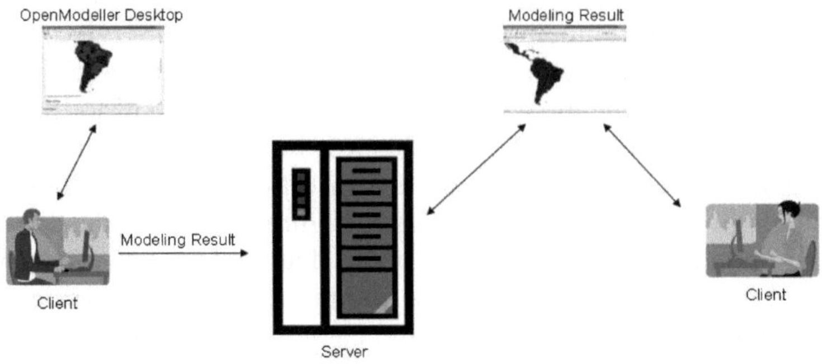

Fig. 6. Experiment Context

The Figure 6 shows the client obtaining the modeling result through the OpenModeller's desktop interface. Then, the client transfers the files with the modeling result to the server. A client can visualize some files of the modeling result through a web interface. These files are: the .html file with the report generated by the openModeller, the xml file with input data and metadata of the modeling algorithm, for instance the data related to species occurrence. The Figures 7, 8 and 9 shows web interfaces which allow visualizing some data results available on the server.

The screen shown in the Figure 7 presents some data of the models inserted into the MySQL database, such as information about the modeling algorithms used by OpenModeller. The form holds links to the files containing data results on the server.

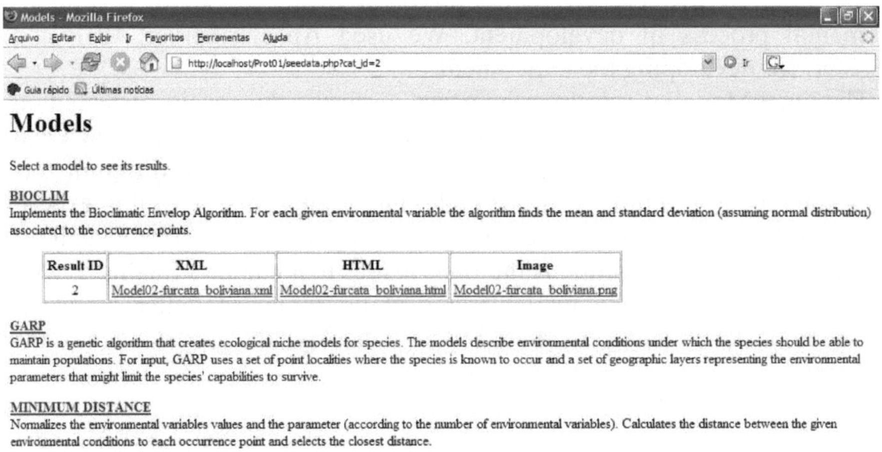

Fig. 7. Database visualization

The client can visualize the files contained in the server by clicking on the respective links. Figure 8 shows the html file containing an openModeller report. This report contains execution information, as the input and output files.

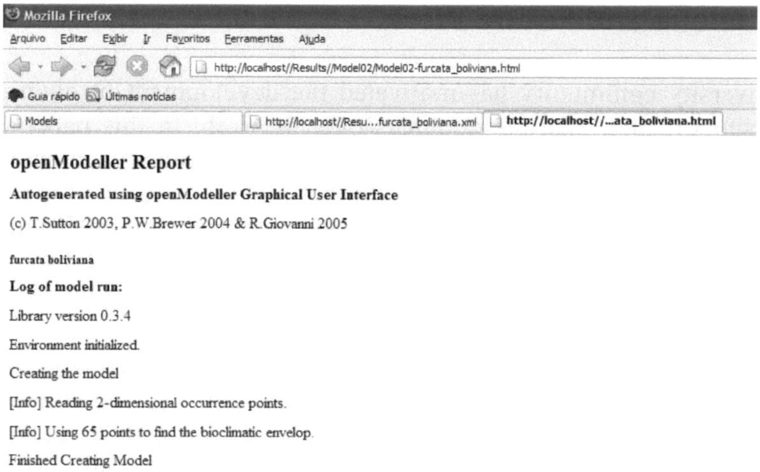

Fig. 8. HTML file visualization

Figure 9 presents the content of an image file containing the modeled area.

Fig. 9. Modeling map file visualization

The map visualized in this file is the species distribution map obtained by the modeling process from OpenModeller. This visualization is one of the applications of WBCMS.

5 Conclusions and Future Work

The Biodiversity community has motivated the development of environments to support their sharing resources over the web. In this paper we presented preliminary ideas about WBCMS - Web Biodiversity Collaborative Modeling Service, a GI Web Service that allows users to cooperate over a species distribution modeling network. The cooperation happens through the cataloguing of models instances. The service allows the knowledge obtained by one modeler or group of modelers to be shared with newcomers and/or other modeler's researchers.

WBCMS is in its initial phase of development. The experiment proposed doesn't contemplate all services that must provided by the proposed architecture. The next step in our work is to define the computational environment, implementing the whole architecture of WBCMS, and perform new experiments with real data and real models and modelers involved.

Acknowledgements

Karla Fook's work is partially funded by FAPEMA[12] (in portuguese: Fundação de Amparo à Pesquisa e ao Desenvolvimento Científico e Tecnológico do Maranhão).

References

[1] Aditya T, Lemmens R (2003) Chaining Distributed GIS Services, International Institute for Geo-Information Science and Earth Observation

[2] Alameh N (2001) Scalable and Extensible Infrastructures for Distributing Interoperable Geographic Information Services on the Internet. Ph.D. thesis, Massachussets Institute of Technology

[3] Alameh N (2003) Chaining geographic information web services. IEEE Internet Computing

[12] http://www.fapema.br/

[4] Anderson G, Moreno-Sanchez R (2003) Building Web-Based Spatial Information Solutions around Open Specifications and Open Source Software. Transactions in GIS 7(4): 447-466.
[5] Bernard L, Einspanier U, Lutz M et al. (2003) Interoperability in GI Service Chains-The Way Forward. 6th AGILE Conference on Geographic Information Science, Muenster
[6] Canhos VP, Souza S, Giovanni RD et al. (2004) Global biodiversity informatics: setting the scene for a "new world" of ecological modeling. Biodiversity Informatics 1: 1-13
[7] CRIA (2005) Projeto speciesLink. from http://splink.cria.org.br/.
[8] CRIA, FAPESP (2005) openModeller: Static Spatial Distribution Modelling Tool. Retrieved agosto/2005, from http://openmodeller.cria.org.br/.
[9] Curbera F, Duftler M, Khalaf R et al. (2002) Unraveling the Web services web: an introduction to SOAP, WSDL, and UDDI. IEEE Internet Computing.
[10] DÄoring M, Giovanni RD (2004) GBIF Data Access and Database Interoperability: A united protocol for search and retrieval of distributed data, CRIA - Centro de Referência em Informação Ambiental
[11] Di L, Chen A, Yang W et al. (2003) The Integration of Grid Technology with OGC Web Services (OWS) in NWGISS for NASA EOS Data. HPDC12 (Twelfth IEEE International Symposium on High-Performance Distributed Computing) & GGF8 (The Eighth Global Grid Forum), Seattle, Washington, USA
[12] Emmott S (2004) Biodiversity: The need for a joint Industry, Governments & Scientific community response. Converging Sciences Conference. Trento, Italy
[13] Foster I, Kesselman C (1999) Computational Grids. The Grid: Blueprint for a New Computing Infrastructure, Morgan-Kaufman
[14] Foster I, Vöckler J, Wilde M et al. (2003) The Virtual Data Grid: A New Model and Architecture for Data-Intensive Collaboration. First Biennial Conference on Innovative Data Systems Research, Asilomar, California
[15] Gibotti FR, Câmara G, Nogueira RA (2005) GeoDiscover – a specialized search engine to discover geospatial data in the Web. GeoInfo 2005 – VII Brazilian Symposium on GeoInformatics, Campos do Jordão, SP, Brazil
[16] Guralnick R, Neufeld D (2005) Challenges building online gis services to support global biodiversity mapping and analysis: lessons from the Mountain and plains database and informatics project. Biodiversity Informatics 2: 56-69
[17] Hall P (2004) Biodiversity E-tools to Protect our Natural World. Converging Sciences Conference. Trento, Italy
[18] Hobern D, Saarenmaa H (2005). GBIF Data Portal Strategy, GBIF.
[19] Osthoff C, Almeida RA, Monteiro ACV et al. (2004) MODGRID – Um ambiente na WEB para desenvolvimento e execução de modelos espaciais em um ambiente de Grades Computacionais. Petrópolis, LNCC - Laboratório Nacional de Computação Científica
[20] Panatkool A, Laoveeraku S (2002) Decentralized GIS Web Services on Grid. Open source GIS - GRASS users conference, Trento, Italy

[21] Pinto GRB, Medeiros SPJ, Souza JMd et al. (2003) Spatial data integration in a collaborative design framework." Communications of the ACM 46(3): 86-90

[22] Stockwell DRB, Beach JH, Stewart A et al. (2006) The use of the GARP genetic algorithm and Internet grid computing in the Lifemapper world atlas o species biodiversity. Ecological Modelling 195(1-2): 139-145

[23] Tsou MH, Buttenfield BP (2002) A Dynamic Architecture for Distributing Geographic Information Services. Transactions in GIS 6(4): 355-381

[24] White R (2004) Helping biodiversity researchers to do their work: collaborative e-Science and virtual organisations. Converging Sciences Conference. Trento, Italy

[25] Zhao Y, Wilde M, Foster I et al. (2004) Grid middleware services for virtual data discovery, composition, and integration 2nd workshop on Middleware for Grid Computing Toronto, Ontario. Canada, ACM Press

Evaluation of OGC Web Services for Local Spatial Data Infrastructures and for the Development of Clients for Geographic Information Systems

Leonardo Lacerda Alves, Clodoveu A. Davis Jr.

Instituto de Informática
Pontifícia Universidade Católica de Minas Gerais

1 Introduction

Interoperability is one of the most important challenges related to GIS. Through the last years, research on interoperability has evolved from the simple off-line exchange of standardized-format files, through the establishment of spatial data clearinghouses, and to the first initiatives in the treatment of semantic aspects of data. Practical interoperability, however, is still hampered by the need to agree on standards, and to develop appropriate tools and methods.

The Open Geospatial Consortium (OGC) has proposed a number of standards to that respect, with the intention of promoting interoperability through the use of services [25]. However, OGC's definition of Web-based services for spatial information predates the World-Wide Web Consortium's definition of the Web service architecture [34]. The necessary adjustments between OGC's and W3C's proposals are still under way.

Meanwhile, we observe that a number of difficulties arise when someone tries to effectively implement the interoperability-through-services approach. Issues regarding fault tolerance, server-independent implementation, delayed-time transactions, privacy, and others reflect the need for further study and discussion. In fact, studies about the conformance of OGC standards to the distributed systems development are still scarce.

This paper extends the work of Alves and Davis Jr [2] and it discusses the current status of service-oriented architectures as applied to interoperable GIS, or, more specifically, to the implementation of local spatial data infrastructures (LSDI). Most existing spatial data infrastructures refer to regional or country-wide data, while LSDI deals with a more complex and rich set of geographic data [24, 28]. Thus, LSDI involve a wide variety of services, while also dealing with users of a wide range of devices, such as PDAs, cell phones and personal computers.

For this discussion, we defined a real-world use case based on an urban context, and developed a services prototype following OGC's abstract model. We used this prototype to assess the engineering guidelines for the server and client development, according to the viewpoints established by the OGC Reference Model [25]. To solve some of the limitations and issues that we have identified, we proposed and developed special infrastructure services which illustrate some deficiencies of OGC specifications. Nevertheless, we do not imply here that such services should become part of the standard. Using the proposed services, it is possible to enable asynchronous communication between clients and services, to access data provided by clients, and to improve on critical points of the services-oriented architecture, such as recovery from failures, dynamic service chains creation, and others, while staying within the OGC Reference Model.

This paper is organized as follows. Section 2 presents concepts about Spatial Data Infrastructures (SDI), in general, and Local Spatial Data Infrastructures (LSDI), in particular, including the ideas behind configuring SDI and LSDI as services. Section 3 introduces our discussion as to the required functionality of a local SDI, and the way to achieve that using the aforementioned infrastructure services. Finally, Section 4 presents our conclusions and indicates some research directions from the concerns presented here.

2 Related Work

2.1 Spatial Data Infrastructures

Spatial Data Infrastructures (SDI) constitute a set of policies, technologies and standards that interconnect a community of spatial information users and related support activities for production and management of geographic information [26]. The idea behind SDI involves avoiding redundant effort and reducing production costs for new and existent datasets through the sharing of resources. In order to achieve this, it is very important that the various partners have convergent interests, agree on common rules and are allowed to make use of data or information produced by others.

Geographic information from one or several partners can be consolidated and thus form important resources for high-level decision-makers. In this case, SDI can be seen as a set of building blocks, as defined by Rajabifard et al [28], in which SDI hierarchies are built through the exchange and consolidation of information from corporate and local levels, to regional and global levels. In this hierarchy, lower levels provide detailed information that helps in the consolidation of the upper, more general, levels [27, 17, 21].

Guiding the technology standardization and, consequently, defining the key elements for spatial data infrastructures, a number of standards were proposed by the OGC, through a framework called OGC Reference Model [25]. This framework has been implemented successfully in state, national and regional scales [12, 13, 30], but reference cases of local SDI are still scarce. Furthermore, most of them are not OGC-compliant [18, 22].

Local SDI has the potential of bringing a numerous group of users together, each of which with distinct needs [9]. This characteristic is in part responsible for an increased level of complexity in local SDI development and deployment [5, 23]. Indeed, LSDI is valuable to all of the other SDI levels as a detailed information source, and the implementation of actual data sources in the local level should evolve simultaneously with, and guided by, the new demands for Geographic Information (GI) from a new range of users, such as business travelers, citizens, small companies and others [23].

SDI can be implemented by chaining services of different sources [1] and integrating software components [15] that can be found in geoportals [20]. The emphasis on services has increased since the emergence of Web services and of the Service-Oriented Architecture (SOA) in the field of distributed systems development. This led Bernard and Craglia [4] to propose a new translation for the the SDI acronym, to mean *Service-Driven Infrastructures*.

2.2 Web Services and OGC Services

Web services are usually seen as Web-based enterprise-wide or inter-organizational applications that use open standards (mostly based on XML) and transport protocols to exchange data with clients, thus forming a loosely-coupled information systems architecture [11, 14]. Web services, as specified by the World Wide Web Consortium (W3C) [34], use the Hypertext Transfer Protocol (HTTP) in the application layer. This protocol only allows the client to perform synchronous calls. This characteristic represents a problem when dealing with delayed-time transactions and call resume.

Some initiatives try to emulate asynchronous calls in Web services through the use of *listeners* that receive responses and forward them to the requester [29], usually over non-HTTP protocols [7]. Other strategies involve the use of the electronic-mail protocol, SMTP [8]. However, as far as we know, no pattern enables asynchronous communication using only Web service standards.

Regarding geographic information specifically, the OGC Web Services specification (OWS) [33] describe a set of functionalities and components which support standardized information exchange. Among the OGC services, there are asynchronous communications only for sensor services [6] such as the Web Notification Service, but the communication protocol is not specified in the standard.

In this work, we are particularly interested in LSDI, i.e., a network of clients and services that deal with an urban context and typical urban applications. The urban-specific OGC proposed platform is called Open Location Services (OpenLS) [19], in which services such as directory (yellow pages service), routing services, geocoding services, and others are grouped. A proposal for urban services of interest as LSDI elements can be found in [10]. The variety of possibilities for using such urban services to build compact and useful geographic information-based applications is staggering. The next section presents one of them.

2.3 Use Case: a Consumer Travel Assistance Application

The OGC Reference Model [25] defines a scenario entitled ``consumer travel assistance" [p. 83] through which a GI consumer uses a mobile client to (1) get its actual position or location; (2) get the destination address, given a telephone number; (3) get a location, given the destination address; (4) get a route between the origin and the destination; (5) determine the traffic, weather and road conditions along the way; and (6) obtain real-time travel advice. Of course, such a task potentially requires a large amount of geographic data, possibly from various sources. This is something that is well above the capabilities of a typical mobile client, thus requiring a new computational model. In this paper, we extend this scenario and use it to implement OpenLS platform clients and an OWS framework, in order to evaluate the OGC Reference Model for LSDI client development.

Our modified scenario involves two basemap services, one routing service, one geocoding service, one directory service for telephone numbers, one directory service for companies, one public transportation service, one public-services service, two emergency services, one gateway for personal location, and two kinds of clients (thin and thick).

While designing LSDI clients to deal with such a variety of services, we have identified some limitations of OGC specifications, more specifically in important engineering requisites such as service availability assurance, recovery in case of failures, privacy control, communication cost reduction and transparent support for the peculiarities of different kinds of client devices.

Even though we have not yet fully developed this scenario as to its actual functionality, we have worked on determining and implementing the ideal communications protocols for each situation, so that some of the limitations inherent to OGC's architecture can be adequately solved beforehand. A prototype was implemented, based mostly on Web service aspects, and leaving the many related GIS issues for the near future. The goal of this prototype is to make sure the required transactions are adequately designed, and all communications issues are solved. The next section presents and discusses the alternatives we have conceived and implemented to that effect.

3 Innovative Services for LSDI

Given the anticipated needs of urban GI applications and the main limitations of GI Web services, some novel infrastructure services are useful to improve the LSDI clients development and the LSDIs themselves.

In the next three subsections, we present the infrastructure services we conceived and developed. The first, called *Data Exchange Service*, is used for persistence maintenance between clients and servers. The second, called *Client Access Service*, is used for information exchange and access control to clients by servers and other clients. The third, called *Transaction Control Service*, is used by clients and servers to improve engineering capabilities such as failure recovery, dynamic service chains creation, workflow definition, and others. The fourth subsection presents details about the prototype implementation and its analysis.

3.1 Data Exchange Service

The Data Exchange Service (DXS) supports interactions between clients and servers (client to server communication), different servers (server to server communication), and different clients (client to client communication) as a workspace where data may be freely stored and retrieved. The objective of this service is to reduce the volume of interactions between clients and servers, and to minimize the connection costs in service invocations and data retrieval, even when failures occur.

Figure 1 shows an example of the DXS in action. A service A, replicated in A' and A" for redundant availability, as it is invoked by a client. Initially, the request parameters are sent to a DXS (1). Next, the ordering of the procedures is defined by the client, in the form of a workflow (2). An invocation process (3) reaches service A, or resorts to services A' or A" in case of failure. In any case, the parameters are recovered from the DXS (4), and the results are stored in that server for future retrieval by another service, or by the client (5).

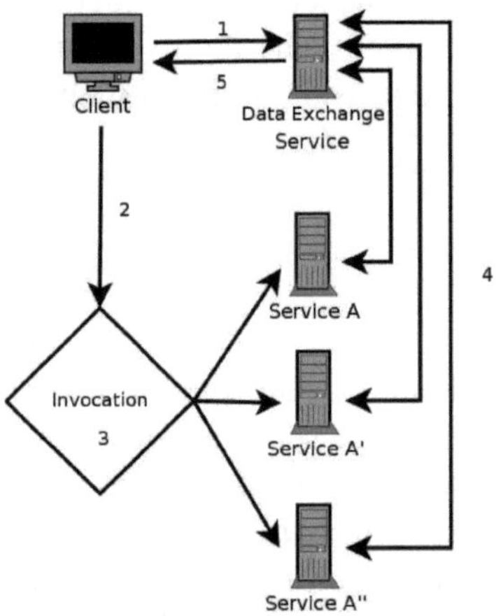

Fig. 1. Data Exchange Service in the invocation of a replicated service

The Data Exchange Service works in three different ways. The first way effectively establishes *asynchronous communication* between the client and the target service. The DXS mediates the communication, so that the client does not have to wait online for a response from the target service. When the processing is through, the client receives a message from the DXS, and is then allowed to retrieve the results.

The DXS also works as a *temporary repository* for intermediate responses in a service chain. Intermediate results are kept by the DXS for the benefit of services along the chain, but the client is only allowed access to the final results, as they become ready.

Furthermore, the DXS supports *failure recovery*, since it can keep information on the status of a service chain, along with intermediate results. With this, recovery and continuation from the client side is possible by re-invoking any service that fails, using the stored parameters, thus avoiding reinitialization of the entire chain.

To understand and to specify how the Data Exchange Service works, we defined the client and service invocation patterns using an UML (Unified Modeling Language) sequence diagram. Figure 2 presents the diagram.

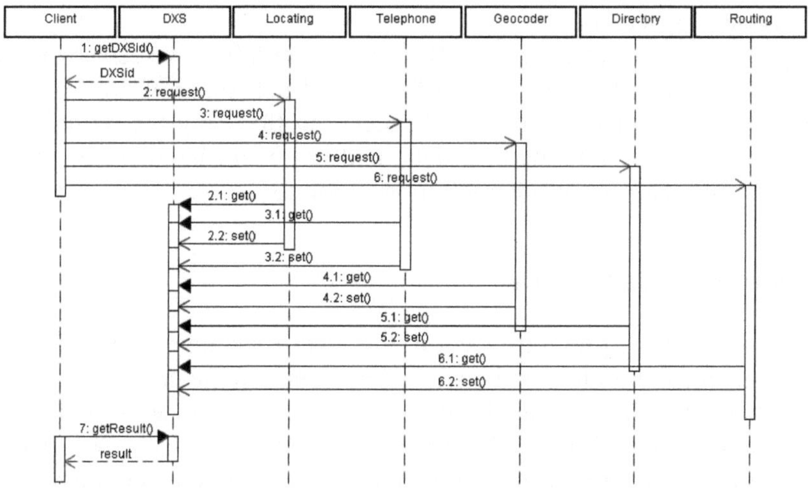

Fig. 2. UML Sequence Diagram for a System using DXS

Messages 1 and 7, between the client and the DXS, are the same generic processes 1 and 5 shown in figure 1, which initiate the processing and collect the final results, respectively.

Messages 2, 3, 4, 5, and 6 invocate all needed services at the beginning of the process, when the client stops and the services keep running. Next, as the *Locating* and *Telephone* services can be executed in parallel, messages 2.1, 2.2, 3.1, and 3.2 can occur in any order.

Finally, messages between 4.1 and 6.2 occur sequentially. Notice that all intermediate data exchange would pass through the client if the DXS was absent. However, using this service, such intermediate response traffic is avoided. Most of the service chain can be invoked in parallel to reduce client work, and the responsibility for most invocation callbacks is transferred to the DXS.

In the highest conceptual level, the Data Exchange Service has thus the function of intercepting service responses and forwarding them to other services (including other DXS) or to the client. In this sense, the communication costs for thick and rich clients remains the same, while costs for thin clients fall drastically. The length of the lifeline remains the same, regardless of the involvement or not of the DXS. Therefore, the DXS offers advantages only when the client has energy limitations or it is slower than the DXS server.

It would be possible to implement DXS-like persistence through simple Web services, but a standardized interface that functions as an infrastructure service is important to ensure the independence between clients and OGC-services providers.

As presented, the DXS may suffer from a number of security issues, such as unauthorized access to data from third parties. However, protocol enhancements can ensure authorized access, so that only the participating services can retrieve intermediate data, and not even the client is allowed access to privileged or confidential information [3].

3.2 Client Access Service

As mentioned in the previous section, the Data Exchange Service enables asynchronous communication between the client and the target service, notifying the client when the results are ready to use. However, asynchronous calls are not supported in W3C Web services using HTTP [7, 29]. Among OGC services, only the Web Notification Service implements asynchronous answers, even though it also does not use a HTTP interface [6].

```
transactionID = service.do(params);
```

Code. 1. A call that return a transaction ID

In an urban context, asynchronous services are often necessary, working as delayed-time transactions for long service chains. They are also necessary in particular applications, such as using servers as sensors to send information to clients without counting on connection-oriented communication.

```
while (!service.isReady(transactionID))
    self.sleep(1000);
result = service.get(transactionID);
```

Code. 2. Getting the service result

Even though normally a client does not have a valid IP (Internet Protocol) address, it is able to access HTTP resources. However, without an IP address it is inaccessible from other clients or services, receiving data only during its request connections. We propose an alternative to reach clients without an IP address. A client can invoke a service using a method that returns a transaction Id, but not the final result. This is illustrated in code 1.

```
transactionID =
    service.do(responseMethod, responseURI,
                                    params)
```

Code. 3. The client define a method for receive a notification when the result is ready

In a second moment, the client invokes the service again through a method that returns the final results, giving the transaction Id, as illustrated in code 2.

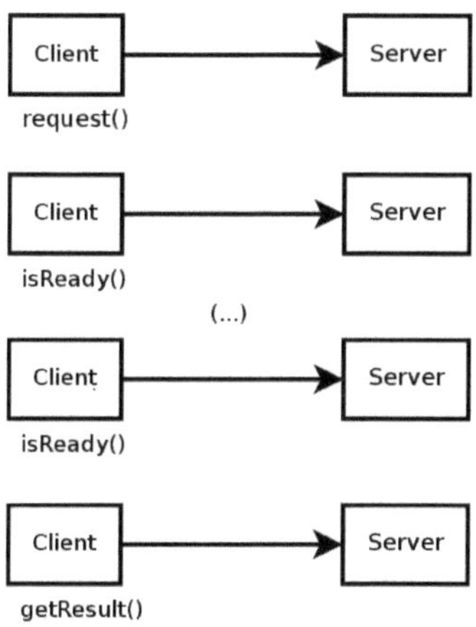

Fig. 3. Client without valid IP address using communication based on HTTP

However, getting results is only possible when the processing has been concluded, and the client uses up network and processing resources to poll the service for a response. Figure 3 presents this communication pattern.

If the client has a valid IP address and implements the functionalities of HTTP servers, we have a different situation. The client waits for responses through a DXS-like service, precisely until it receives a response from the service, informing that the results are ready. Code 3 illustrates a call that specifies a response method (HTTP, SMTP, SMS, etc) as part of its parameters, along with an URI (a address through which the client expects the notification when the process had ends).

```
while (self.waiting(transactionID)) {
    // do nothing
}
result = service.get(transactionID);
```

Code. 4. Waiting for Notification from the Gateway or Other Service

Figure 4 shows the described communication protocol, and the code 4 illustrates the last call from the client to the service when de service processing becomes ready.

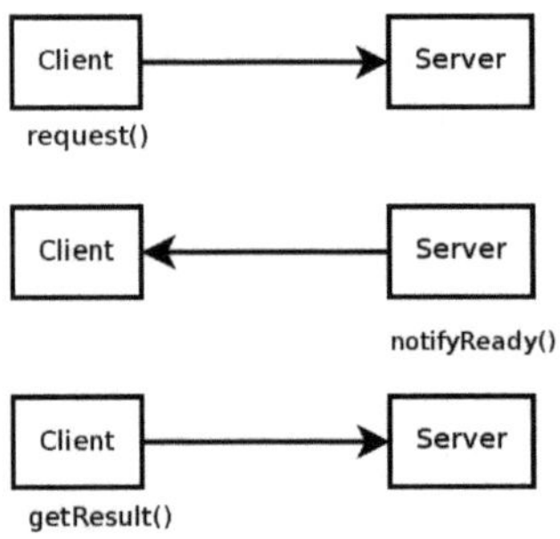

Fig. 4. Client with valid IP address using communication based on HTTP

The last discussed pattern is dependent on a local gateway service such as a DXS, and it is adopted when the client does not have a valid IP address, or when the client is protected behind a firewall, or when the client does not want to be identified. In this case, the client can access the target service directly or indirectly, but the notification process always occurs between the chosen gateway and the client, as illustrated in figure 5. The client implements the same code presented in code 3 and in code 4. We call this approach the *Client Access Service* (CAS).

The Client Access Service offers means to use clients as sensors for others, and means to enable the client to work as a Data Exchange Service provider (specially in the case of thick clients). In both cases, the client is able to provide some information to other clients and services and to support a number of concurrent processes, with adequate security and privacy constraints.

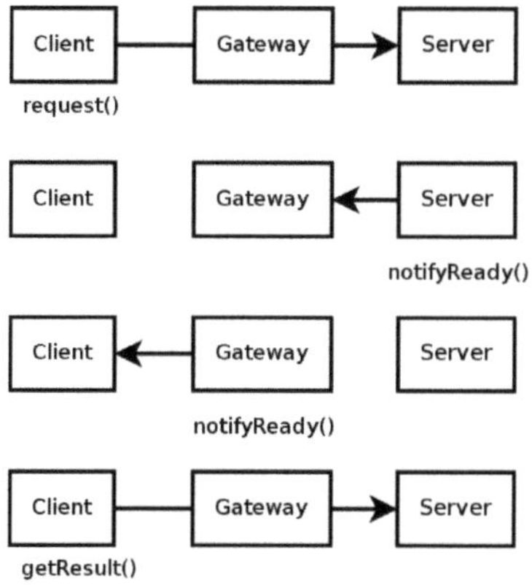

Fig. 5. Client without valid IP address using a Gateway as mediator

3.3 Transaction Control Service

Through the two services previously presented, a GI client can improve its processing capabilities and make the access to geographic information easier. However, it is hard to develop a generic client, suitable to different LSDIs. Furthermore, some capabilities, such as fault tolerance and engineering transparency, are actually the server's responsibility, but shifting this responsibility towards the client is sometimes convenient, specially for mobile clients in an urban context.

When traveling, the client should be independent of specific service chains. In this case, when a client finds itself in a distant city, it can load and execute a chain of different services to obtain local information.To do this, services can be readily available within chains, but it would be better if each service was available individually and listed in a public catalog service. The selection and chaining of services can occur dynamically whenever the client needs to access them.

The Transaction Control Service (TCS) performs transformations in generic chaining code in order to generate a fully functional service chain, suited to a particular context. The generic code defines a basic workflow that is to be followed by the client during the execution process. That same basic workflow can serve as a template for other service chains, counting on the Data Exchange Service for persistence and to store the processing state.

The codes 5 and 6 present the result of using the TCS for the conversion of an abstract and generic codes into client-specific service chains in the particular context of a routing application. The main objective of this service is to ensure that neither client nor service development are dependent on the local context, i.e., on peculiarities of the services that are available at each different location. The code 5 is presented as a pseudocode, albeit the code is actually specified using XML.

```
// Client software in control
target = client.from();
myLocation = client.from();

// Concurrent commands without user interaction
myLoc = location(myLocation,Tdirectory) |
p=pointsOfInterest(location(target),Tdirectory);

// Client software in control
client.to(p);
p2 = client.from();

// Non-interactive commands
query(Troute,myLoc, p2, location(target));
```

Code. 5. Abstract code, independent of technology and services

The generic code makes changes on the previous code when the Data Exchange Service and the geographic services are introduced in the algorithm and the requesites of the current LSDI are taken into account.

```
// Concurrent commands without user interaction
myLocAd = cityHall.location(myLocation,
          Tdirectory, dxs);

cityHall.location(target, dxs.setItem(0));
teleCo.location(target, dxs.setItem(0));
coAssociation.location(target, dxs.setItem(0));
tourismGuide.location(target, dxs.setItem(0));

pAd = cityHall.pointsOfInterest(dxs.getId(0),
      Tdirectory, dxs);
```

Code. 6. Part of specific code adapted to a LSDI and thin client

Finally, the code 6 should implement an algorithm as efficient as possible to the client software, which may be designed according to the needs of specific clients and users for the current LSDI.

With these services, LSDI clients can be implemented based on informational and computational aspects, avoiding the introduction of technological aspects into client code. Thus, thick-, rich-, and thin-clients can be implemented transparently, and their particular constraints are addressed by a generic workflow that changes easily when the client moves to another place or has its capacity improved.

3.4 Prototype Implementation and Analysis

We have implemented Web services for GIS as specified by the OGC, following the Abstract Model specification. The distribution patterns are (a) 1 Client to 1 Server (or known provider) to *n* Servers, where there is a provider which mediates resources from others, and (b) 1 Client to *n* Servers, where the client performs all of the tasks related to dealing access to Web services and geographic information processing.

The first pattern (a) made it easier to implement clients and servers with important engineering requisites, but introduces much dependency on known providers, in situations where no information about their resources and about external resources is available.

However, to enable client access to information about available resources, it is necessary to abandon the "known provider" role. This led us to implement the pattern (b), 1 Client to *n* Servers. This pattern increases the costs of processing and traffic of intermediate data, because the client requests data to several services, and once it receives their responses it usually performs some operations and submits a large volume of data to other services in the service chain. However, this pattern enabled us to perceive and to define some engineering requisites on the client side.

Next, we separated design issues from technologic ones, and we grouped them into three groups, which constitute the services presented in the previous subsections.

By executing the features grouped into the Data Exchange Service, the intermediate data traffic in non-interactive processing was avoided on the client. In addition, in case of failure no previously processed data was lost. Nevertheless, we noticed no significant effect in interactive processing when there are multiple inputs during the service chain execution.

Through the Client Access Service, the delay and communication costs were reduced by avoiding the use of the network by the client when it needs to poll the service about its readiness before it really becomes ready.

Finally, we created a subset of operations and defined service chains using XML. The Transaction Control Service was then used to convert the XML data, given the client's requisites and characteristics, to a service chain which uses the above infrastructure services, automatically.

4 Conclusion

4.1 Results

This work evaluated the engineering aspects of OWS specifications and the main OGC services in the context of urban applications. Through these applications, we identified some implementation constraints that are characteristic of GI systems, such as non-standardized fault tolerance mechanisms, clients that are strongly dependent of providers, and others.

To achieve this, we implemented a prototype based on the services abstract model for a real-world use case, and tested the usefulness of OGC standards to LSDI, specifically considering LSDI client development. As the objectives of standardization include to guide uniform application deployment and services interoperability, distribution transparencies, such as access, failure, and persistence transparencies, become essential.

Since the availability of client software contributes to LSDI diffusion and implementation, we have also developed new infrastructure services, which facilitate the development of LSDI clients and interoperable GIS with urban characteristics. Thus, the new services constitute a synthesis of missing features in the standard technologies, and this prototype allows the exploration of other issues related to OGC standards and urban GIS applications in future research efforts.

4.2 Main Contributions

Through the Data Exchange Service, clients – either thin, rich, or thick – can be transparently developed without previous distinction. There are additional advantages for thin and rich clients, which have greater limitations of energy, storage, communication, and processing capacity.

The DXS replaces the persistence function of service providers and local storage with a third-party neutral service, through which the services chain exchanges parameters and results of its inner processing.

The establishment of the services chain is improved by two solutions. The first one is focused on the client, while the second one benefits from the communication capabilities among servers.

In the first alternative, the services chain orchestrated by the client defines the providers before and during the invocation of services. Then, the client becomes responsible for fault tolerance and it can choose alternative services according to established parameters. Additionally, when the client travels or in case of emergencies, the whole service set may be changed.

Nevertheless, the services chain may be orchestrated by a W3C- or OGC-like service, that calls others. Normally the client does not participate in service selection for the composition of a chain. It can only refuse a provider by selecting another. However, even this situation benefits from a shift of responsibility from the server to the client. The insertion of the Data Exchange Service into server workflow reduces the reissuing of requests in case of failures (specially if the client is thin), and increases the parallelism level of processing.

Therefore, we proposed a set of new services, which facilitate the development of GI clients for LSDI-based systems, and contribute to improve the diffusion and implementation of LSDIs.

4.3 Future Work

We identify three main directions for future work. First, we discuss the potential use of clients as information providers. Next, we propose some engineering enhancements to the development of services. Finally, we envision the possibility of developing new LSDI services, enhancing the prototype presented in this paper.

The GI client, viewed as a server of parameters for the service chain, can potentially become the server of various kinds of information. Information perceived directly by the user, such as traffic status [31], along with previously collected data, perceived quality of information, ontologies about its interests, geographic position, and others, can be passed along to other users through services. Possible applications to this are better traffic management using vehicles as real time sensors, more precise route planning by information exchange with other GI users, quality-of-service control, among others. For traffic status monitoring, information may be exchanged between these “sensors” by active client-server connections (in the case of wireless sensor networks with limited energy capacity), or peer-to-peer connections (in the case of wireless sensors networks without energy restrictions, i.e. ad-hoc networks constituted by in-transit vehicles), with new application possibilities, such as the dissemination of warnings or instructions. Peer-to-peer connections, when energy is not an issue, should improve the response time of applications [32], because previously collected data of other users is available. Geocoding, routing and locating services can then be adapted for mobile geographic objects, such as vehicles (public transportation, emergency, particular vehicles), or for personal locating (for instance, using cell phones). However, additional studies about these applications are required, in order to deal with concerns such as security and privacy. For effective privacy control, privacy parameters could be dynamically negotiated between clients and servers.

Clients, functioning as information providers, can also be the source of a measure of quality of service. Client feedback can be used to establish perceived quality, and to identify the quality parameters most valued by a user group. Ontologies on perceived quality parameters can then be developed from these observations.

As to engineering, fault tolerance parameters can be defined and negotiated according to availability contracts that should be automatically verified through a search into the service chain and evaluation of common failure points between main and substitute services. This is specially important in emergency response, critical services applications, and others.

Another engineering point regards workflow descriptions. In some cases, it is helpful that services be localized and dynamically injected into the original workflow, as in the case of the routing service, when routes cross regions not covered by the loaded data sources. Therefore, additional improvements should be made to workflow description mechanisms.

Davis Jr and Alves (2005) [10] propose several services for LSDI, considering the specific demands for urban geographic applications. We think that further services related to this level of infrastructure can be proposed and investigated, using the prototype presented here as a basic framework.

Finally, the creation of a more complex LSDI prototype development environment is important to enable observations and experimentation on current and upcoming OGC standards. Such a framework should be based on the OGC's implementation specifications rather than on an abstract specification model, as we did. Thus, it can be used to develop further formal studies about several concerns (privacy, performance, security, and so on) with real-world problems. It would also be important to build such a prototype development environment using free or open-source software, so that the community involved in spatial data infrastructures can benefit, while being able also to contribute to it.

Additionally, we think the standardization process, as it is today, can produce social problems if applied to LSDI. Therefore, according to [16], efforts to adapt to older to new standards can cause fragmentation, complexity and heterogeneity rather than solve this problems. An important research question should be which factors contribute to success and failure of standardization efforts on LSDI.

References

[1] Alameh, N. (2003). Chaining geographic information web services. IEEE Internet Computing, 7(5):22–29.

[2] Alves, L. L. and Davis Junior, C. A. (2006). Interoperability through Web Services: Evaluating OGC Standards in Client Development for Spatial Data Infrastructures. In VIII Brazilian Symposium on Geoinformatics Proceedings, pages 193–208. INPE.

[3] Belussi, A., Bertino, E., and Catania, B. (2004). An authorization model for geographical maps. In GIS'04 Proceedings, pages 82–91. ACM.

[4] Bernard, L. and Craglia, M. (2005). SDI – From Spatial Data Infrastructure to Service Driven Infrastructure. In Research Workshop on Crosslearning between Spatial Data Infrastructures and Information Infrastructures.

[5] Bishop, I. D., Escobar, F. J., Karuppannan, S., Suwarnarat, K., Williamson, I. P., Yates, P. M., and Yaqub, H. W. (2000). Spatial data infrastructures for cities in developing countries: Lessons from the Bangkok experience. Cities, 17(2):85–96.

[6] Botts, M., Robin, A., Davidson, J., and Simonis, I. (2006). Sensor Web Enablement Architecture, OGC document 06-021r1. OGC, Wayland.

[7] Brambilla, M., Ceri, S., Passamani, M., and Riccio, A. (2004). Managing asynchronous web services interactions. In Web Services, 2004. Proceedings. IEEE International Conference on, pages 80–87. IEEE.

[8] Chung, S., Pan, J. R., and Davalos, S. (2006). A special web service mechanism: Asynchronous .NET web services. In Telecommunications, 2006. AICT-ICIW '06. International Conference on Internet and Web Applications and Services/ Advanced International Conference on, pages 212–212. IEEE.

[9] Davis Junior, C. A. (2005). Considerations from the Development of a Local Spatial Data Infrastructure in Brazil. In Research Workshop on Cross-learning between Spatial Data Infrastructures and Information Infrastructures.

[10] Davis Jr, C. A. and Alves, L. L. (2005). Local spatial data infrastructures based on a service-oriented architecture. In VII Brazilian Symposium on Geoinformatics Proceedings, pages 84–89. INPE.

[11] Davis Jr, C. A. and Alves, L. L. (2007). Geospatial Web Services. In Encyclopedia of Geographic Information Systems. Springer-Verlag, Berlin.

[12] Deegree (2006). Deegree Project (WMS, WFS, WCS, iGeoPortal, deeJUMP). Lat/Lon GmbH http://www.deegree.org.

[13] Demis (2006). Demis Web Map Server. Demis http://www.demis.nl/.

[14] Ferris, C. and Farrell, J. (2003). What are web services? Communications of the ACM, 46(6):31.

[15] Granell, C., Gould, M., and Ramos, F. (2005). Service composition for SDIs: Integrated components creation. In: Proceedings of II Internation on Geographic Information Management.

[16] Hanseth, O., Jacucci, E., Grisot, M., Aanestad, M. (2006). Reflexive Standardization: Side effects and complexity in standard making. MIS Quarterly. Vol. 30. Special Issue, pages 563–581.

[17] Jacoby, S., Smith, J., Ting, L., and Williamson, I. (2002). Developing a common spatial data infrastructure between state and local government: An Australian case study. IJGIS, 16(4):305–322.
[18] Kumar, P., Singh, V., and Reddy, D. (2005). Advanced traveler information system for Hyderabad City. Intelligent Transportation Systems, IEEE Transactions on, 6(1):26–37.
[19] Mabrouk, M. (2004). OpenGIS Location Services (OpenLS): Core Services, OGC document 03-006r3. OGC, Wayland.
[20] Maguire, D. J. and Longley, P. A. (2005). The emergence of geoportals and their role in spatial data infrastructures. Computers, Environment and Urban Systems, 29(1):3–14.
[21] Man, W. H. E. D. (2006). Understanding SDI; complexity and institutionalization. IJGIS, 20(3):329–343.
[22] Mansourian, A., Rajabifard, A., Zoej, M. J. V., and Williamson, I. (2006). Using SDI and web-based system to facilitate disaster management. Computers & Geosciences, 32(3):303–315.
[23] Masser, I. (2005). Some priorities for SDI related research. In Proceedings of the FIG Working Week and GSDI 8: From Pharaohs to Geinformatics, Cairo, Egypt, page 11. FIG.
[24] Nedovic-Budic, Z., Feeney, M.-E. F., Rajabifard, A., and Williamson, I. (2004). Are SDIs serving the needs of local planning? case study of Victoria, Australia and Illinois, USA. Computers, Environment and Urban Systems, 28(4):329–351.
[25] Percivall, G. (2003). OGC Reference Model OGC 03-040. OGC, Inc.
[26] Phillips, A., Williamson, I., and Ezigbalike, C. (1999). Spatial data infrastructure concepts. The Australian Surveyor, 44(1):20–28.
[27] Rajabifard, A. and Williamson, I. P. (2001). Spatial data infrastructures: Concept, hierarchy, and future directions. In Geomatics'80, page 11. Tehran, Iran.
[28] Rajabifard, A., Williamson, I. P., Holland, P., and Johnstone, G. (2000). From local to global SDI initiatives: a pyramid of building blocks. In IV Global Spatial Data Infrastructure Conference Proceedings.
[29] Ruth, M., Lin, F., and Tu, S. (2005). A client-side framework enabling callbacks from web services. In Web Services, 2005. ECOWS 2005. Third IEEE European Conference on, page 12. IEEE.
[30] Skylab (2006). J2ME OGC WMS Client. Skylab http :// www . skylabmobilesystems . com / en / products / j2me wms client . html.
[31] Tao, X., Jiang, C., and Han, Y. (2005). Applying SOA to intelligent transportation system. In Proceedings of the 2005 IEEE International Conference on Services Computing, pages 101–104. IEEE.
[32] Wang, H., Zimmermann, R., and Ku,W.-S. (2005). ASPEN: An adaptive spatial peer-to-peer network. In GIS'05 Proceedings, pages 230–239. ACM.
[33] Whiteside, A. (2005). OpenGIS Web Services Common Specification, OGC document 05-008. OGC, Wayland.
[34] World Wide Web Consortium (2004). Web Services Architecture W3C Working Group Note (Feb. 11 2004). W3C. www.w3.org/TR/2004/NOTE-ws-arch-20040211/.

Towards Gazetteer Integration through an Instance-based Thesauri Mapping Approach

Daniela F. Brauner, Marco A. Casanova, Ruy L. Milidiú

Departamento de Informática
Pontifícia Universidade Católica do Rio de Janeiro

1 Introduction

A *gazetteer* is a database that stores information about a set of geographic features, classified using terms taken from a given *feature type thesaurus*. A geographic information system may integrate one or more gazetteers to create a consolidated information source about the data the system stores, for example [9]. However, as in a data-warehouse creation process, gazetteer integration requires aligning feature type thesauri, which is the central question we address in this paper.

Our approach uses a mapping rate estimator that computes weighted relationships between terms of distinct thesauri by pre-processing common instances from pairs of gazetteers. Let *G* and *G'* be the gazetteers to be integrated, and assume that they use thesauri *T* and *T'*, respectively. Quite simply, if we have data about a geographic feature *f* from *G* classified as *t* (a term from *T*) and, again, data about *f* from *G'*, but classified as *t'* (a term from *T'*), then *f* establishes some evidence that *t'* maps into *t*. Note that this strategy depends on the assumption that we can recognize when data from *G* and *G'* represent the same geographic feature or not. In this paper, we use the spatial location to deduce that sets of data from *G* and *G'* indeed represent the same geographic features or not.

As for related work, in the area of mediator construction, we may single out the OBSERVER system [11, 12], which uses multiple ontologies, described in a description logics formalism, to access heterogeneous and distributed data sources. OBSERVER requires that conventional mappings between a data source and the base ontology be manually defined. By contrast, our approach automatically generates weighted mappings, working with thesauri.

In the area of ontology mapping, we may highlight the GLUE system, that uses multiple learning strategies to help find mappings between two ontologies [4], the AnchorPROMPT ontology alignment tool, that automatically identifies semantically similar terms [13], and the Chimaera environment, that provides a tool to merge ontologies based on their structural relationships [10]. These three tools work with fully formalized ontologies and, to a varying extent, depend on user intervention. The CATO tool aligns thesauri using mostly syntactical similarities between terms and the thesauri structure [2].

Our approach differs from such systems in two aspects. First, like CATO, we work only with the terms and their structure (the broader term/narrow term relationship). That is, we do not require a fully formalized terminology, using an ontology language. However, unlike CATO, to align two terms, we draw evidence from the way the gazetteers classify geographic features, not merely from a syntactical similarity between the terms.

Castano et al. (2004) [3] describe the H-Match algorithm to dynamically match ontologies. H-Match provides, for each concept from an ontology, a ranked list of similar concepts from the other ontology. Four matching models are used to dynamically adjust the matching process to different levels of richness of the ontology descriptions. Spertus et al. (2005) [15] evaluate the performance of six similarity measures, used to recommend communities to members of Orkut social network communities, adopting the L2 vector normalization (L2-Norm) measure.

This paper is organized as follows. Section 2 summarizes preliminary definitions. Section 3 contains a motivating example. Section 4 describes our instance-based approach to thesauri mapping, including experimental results. Finally, Section 5 contains the conclusions and directions for future work.

2 Gazetteer and Thesauri

A *thesaurus* is defined as *"a structured and defined list of terms which standardizes words used for indexing"* [16] or, equivalently, "*the vocabulary of a controlled indexing language, formally organized so that a priori relationships between concepts (for example as "broader" and "nar-*

rower") are made explicit" [7]. A thesaurus usually provides: a *preferred term,* defined as the term used to consistently represent a given concept; a *non-preferred term*, defined as a synonym or quasi-synonym of a preferred term; relationships between the terms, such as *narrower term (NT)*, indicating that a term – the *narrower term* – refers to a concept which has a more specific meaning than another term – the *broader term (BT)*.

A *gazetteer* is *"a geographical dictionary (as at the back of an atlas) containing a list of geographic names, together with their geographic locations and other descriptive information"* [17]. For our purposes and omitting details, we consider that a gazetteer is a geographic object catalog, where each object has as attributes:

- a unique *object ID*
- an *object type*, whose value is a term taken from an *object type thesaurus*
- a *name*, which takes a character string as value
- optionally, a *location*, which approximates the object's position on the Earth's surface

For simplicity, we assume that each object has only one type and one name (which is not necessarily a key). We note that geographic objects are often called *geographic features*, or simply *features* [14]. Hence, a gazetteer thesaurus is also referred to as a *feature type thesaurus*.

Let G_A and G_B be gazetteers, with thesauri T_A and T_B, respectively. Supposes that one wants to load data from G_A into G_B. Assume that the gazetteers are *homogeneous* in the sense that, given any two features, f_a and f_b, from G_A and G_B, it is possible to detect when f_a and f_b denote the same real world object. This is more an assumption than a definition since we leave it open what is the exact procedure used to detect identical objects.

We are interested in reclassifying features from G_A using the feature type thesaurus of G_B. That is, we want to map the feature types from T_A into feature types from T_B in such a way as to preserve the intended classification scheme of T_A as much as possible.

3 A Motivating Example

As a motivating example, we use two gazetteers that are available over the Web, the GEOnet Names Server and the Alexandria Digital Library Gazetteer.

The GEOnet Names Server (GNS) [5] provides access to the National Geospatial-Intelligence Agency (NGA) and the U.S. BGN database of foreign geographic names, and contains about 4 million features with 5.5 million names. The Alexandria Digital Library (ADL) Project [1, 6] is a research program to model, prototype, and evaluate digital library architectures, gazetteer applications, educational applications, and software components. The ADL Gazetteer has approximately 5.9 million geographic names, classified according to the ADL Feature Type Thesaurus (FTT).

In what follows, we will refer to the ADL Gazetteer and the GEOnet Names Server as G_A and G_B, respectively, and to their thesauri as T_A and T_B. We will consider only countries and cities in the examples that follow.

We now briefly discuss how an instance-based technique may induce a mapping from T_A into T_B thereby enabling the reclassification of the instances migrated from G_A to G_B.

Table 3.1 illustrates how T_A and T_B classify features differently. The second and third columns show how G_A and G_B classify the countries and cities listed in the first column. For example, G_A classifies 'Brazil' as 'Countries', whereas G_B classifies 'Brazil' as 'PCLI'.

Table 3.1. Results of querying countries and cities in the ADL Gazetteer and the GEOnet Names Server

Entry name	ADL Gazetteer (T_A)	GEOnet (T_B)
Brazil	Countries	PCLI
Canada	Countries	PCLI
Germany	Countries	PCLI
Italy	Countries	PCLI
Belgium	Countries	PCLI
Scotland – UK	AdministrativeArea	AREA
Wales – UK	AdministrativeArea	AREA
Rio de Janeiro – Brazil	Populated Places	PPLA
São Paulo – Brazil	Populated Places	PPL
Rome – Italy	Capitals	PPLC
Brussels – Belgium	Capitals	PPLC

In fact, all 5 entries in Table 3.1 that G_A classifies as 'Countries', G_B classifies as 'PCLI'. Therefore, if we would like to load G_A into G_B, this small sample provides us with evidence that instances from G_A classified as the term 'Countries' from T_A have to be loaded into G_B reclassified as the term 'PCLI' from T_B. Moreover, this small sample does not exhibit any conflicting classifications.

4 Instance-based Thesauri Mapping Approach

4.1 Mapping Rate Estimation Model

Our goal is to integrate gazetteers that may use different thesauri to classify their features. To solve vocabulary conflicts, we focus on estimating weighted relationships between terms of distinct thesauri. To achieve this goal, we propose to collect statistics about the common instances from both gazetteers.

Consider two gazetteers, G_A and G_B, and assume that they adopt thesauri T_A and T_B, respectively. Suppose also that we are interested in mapping terms from T_A to T_B.

We say that features f_a in G_A and f_b in G_B, respectively, are *equivalent*, denoted $f_a \equiv f_b$, when they represent the same (real-world) object. In this case, we also say that t_a and t_b *map to each other*, where $t_a \in T_A$ and $t_b \in T_B$ are the types of f_a and f_b, respectively.

The exact procedure that computes instance equivalence depends on the application, as indicated at the end of Section 2. However note that, in the geographic domain, we have various geo-referencing schemes that associate each geographic feature with a description of its location on the Earth's surface. This location acts as a universal identifier for the feature, or at least an approximation thereof. In our approach, we use the feature location to detect equivalent instances and to count the frequency of pairs of terms from different gazetteer thesauri. In other words, we analyze which entries from different gazetteers represent the same geographic object and then calculate a similarity measure between their respective types.

In detail, we define F_A as the set of all $f_a \in G_A$ such that there is $f_b \in G_B$ such that $f_a \equiv f_b$ (and similarly for $F_B \subseteq G_B$). Assume that we have already computed F_A and F_B. We use F_A and F_B to estimate $n(t_a)$, $n(t_a,t_b)$ and $P(t_a,t_b)$ as follows:

1. Compare the features in F_A with those in F_B to compute $n(t_a,t_b)$, defined as the sum of the occurrences of pairs of objects f_a and f_b such that:
   - $f_a \in G_A$ and $f_b \in G_B$
   - $f_a \equiv f_b$
   - t_a and t_b are the types of f_a, and f_b, respectively

2. Compute $n(t_a)$, the number of occurrences of objects $f_a \in F_A$ such that t_a is the type of f_a.

3. Compute $P(t_a,t_b)$ using Eq. (4.1), an estimation for the frequency that the term t_a maps to the term t_b, for each pair of terms $t_a \in T_A$ and $t_b \in T_B$. We call $P(t_a,t_b)$ the *mapping rate estimator* for t_a and t_b.

$$P(t_a,t_b) = \frac{n(t_a,t_b)}{n(t_a)} \tag{4.1}$$

Note that, the above procedure is symmetric in t_a and t_b. Hence, the entire process can be easily adapted to compute estimations for the frequency that terms in T_B map into terms in T_A. Indeed, it suffices to compute $n(t_b)$, instead of $n(t_a)$ and change the denominator of Eq. (4.1) to $n(t_b)$.

4.2 Experiments with Geographic Data

In order to illustrate the mapping rate estimation model proposed in Section 4.1, we present results using the ADL Gazetteer (G_A) and the GEOnet Names Server (G_B). Section 4.2.1 describes how the data were obtained from these gazetteers. Section 4.2.2 discusses how the model was validated and calibrated. Section 4.2.3 contains the test results.

4.2.1 Data Collection

To facilitate the training step, data were collected from the gazetteers servers and stored locally. G_A was consulted using version 1.2 of the ADL Gazetteer Service Protocol, an XML- and HTTP-based protocol for accessing the ADL Gazetteer [8]. Several queries where submitted to G_A, restricted to the Brazilian geographic area, retrieving 16,783 registries in the standard ADL report format (in XML). The returned XML was parsed and the registries were stored in a relational database. As for G_B, data were downloaded from the GEOnet Names Server Web site, which contains files with information about geographic names. The downloaded Brazilian file had 87,608 registries. The available data were partitioned into a tuning set and a testing set, used to tune and to test the model, respectively.

The ADL Feature Type Thesaurus (FTT) has 1,262 terms, organized hierarchically and related using an extended set of the basic thesaurus relationships, as presented in Table 4.1. An example including the list of the ADL FTT top terms is shown in Table 4.2. For this experiment, we consider that the size of T_A is the number of preferred terms (210 terms). The GEOnet thesaurus (T_B) has 642 terms, organized under a single category level with 9 top terms (Table 4.2). The GEOnet thesaurus includes the term code, name, and a textual description.

Table 4.1. ADL Feature Type Thesaurus relationships

Abbreviation	Relationship Name
USE	Use
UF	Used for
USW	Used with
UFW	Used for with
BT	Broader term
NT	Narrower term
RT	Related term
SN	Scope note
DF	Definition
HN	History note

Table 4.2. Top terms from ADL FTT and GEOnet thesaurus

ADL FTT top terms	GEOnet thesaurus top terms
Administrative Areas	Populated Place
Hydrographic Features	Administrative Region
Land Parcels	Area
Manmade Features	Vegetation
Physiographic Features	Streets/Highways/Roads
Regions	Hypsographic
	Hydrographic
	Undersea
	Spot Features

4.2.2 Model Evaluation

To validate the mapping rate estimation model, the data collected was partitioned into two disjoint datasets: the tuning set and the testing set. The tuning set was in turn partitioned into six sets, and each set was partitioned into training set and validation set to apply the 6-fold cross-validation method to estimate the accuracy and recall of the model, and to discover the *threshold mapping rate.*

Table 4.3 shows the six tuning sets and their subsets: the training (T_k) and the validation (V_k) sets with the number of pairs of terms covered. The validation sets were manually labeled with *True* or *False* for each occurrence of pairs of terms. Pairs labeled with *True* indicate that the terms indeed map to each other. The labeling was made by comparing thesauri descriptions and a brief check of equivalent entries, with the help of a geographic domain expert.

Table 4.3. Tuning sets for 6-fold cross-validation technique

Dataset.Id	Dataset	Pairs
V_1	Ex_1	92
T_1	Ex_2, Ex_3, Ex_4, Ex_5, Ex_6	180
V_2	Ex_2	87
T_2	Ex_1, Ex_3, Ex_4, Ex_5, Ex_6	189
V_3	Ex_3	67
T_3	Ex_1, Ex_2, Ex_4, Ex_5, Ex_6	197
V_4	Ex_4	46
T_4	Ex_1, Ex_2, Ex_3, Ex_5, Ex_6	191
V_5	Ex_5	68
T_5	Ex_1, Ex_2, Ex_3, Ex_4, Ex_6	183
V_6	Ex_6	78
T_6	Ex_1, Ex_2, Ex_3, Ex_4, Ex_5	174

In the *k*-fold cross-validation method the model is trained and tested *k* times, using T_k for training and V_k for validation.

The *threshold mapping rate* is the value above which the mapping rates $P(t_a,t_b)$ are considered. The mapping rates of the pairs of terms of the training sets were estimated several times, varying the threshold value from 0 to 1, by 0.1, to discover the best value (see Figure 4.1), i.e., the threshold value with which the model obtained the best accuracy value. The cross-validation process compares these results with the labeled pairs from each validation set. Figure 4.1 shows that the best results were obtained with threshold mapping rate equal to 0.4 (with respect to the cross validation from T_A to T_B).

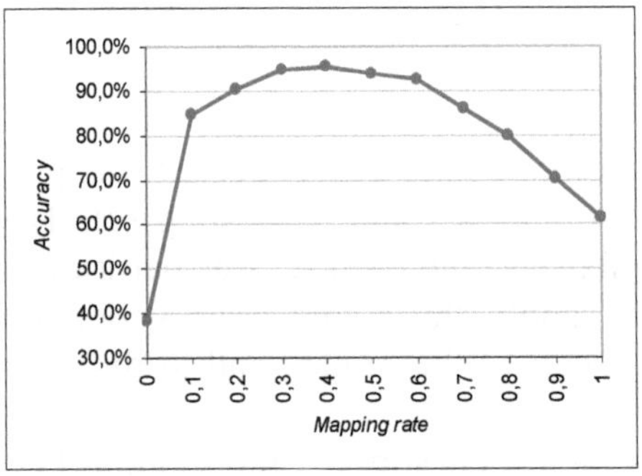

Fig. 4.1. 6-fold cross-validation results from T_A to T_B

4.2.3 Test

To test the mapping rate estimation model, we use the testing set and the threshold mapping rate 0.4 estimated during the model evaluation.

In the cross-validation technique, the *accuracy* is estimated as the number of pairs that are correctly matched, when checking the training sets against the validation sets, divided by the overall number of estimated mappings, with respect to the training sets. The *recall* is estimated as the number of pairs that are correctly matched, divided by the number of labeled pairs, with respect to the validation sets.

As a result of the test step, we have 26 pairs of terms aligned with mapping rate greater than 0.4, from T_A to T_B, with accuracy of 89.7% and recall 81.3%.

Table 4.4. Aligned terms during test step

t_a	t_b	$P(t_a,t_b)$
airport features	AIRF	0.58346
historical sites	RUIN	0.80031
rivers	STMA	0.97371
waterfalls	FLLS	0.92952

Table 4.4 shows examples of the aligned terms. For example, 'waterfalls' from T_A aligns with 'FLLS' from T_B with mapping rate '0.92952'. These values indicate that features migrated from G_A into G_B, formerly classified as 'waterfalls', will be reclassified as 'FLLS' from T_B.

5 Conclusions

In this paper, we addressed the question of thesauri alignment in gazetteer integration, using an instance-based thesauri mapping approach to reclassify the objects loaded from one gazetteer to the other.

Our approach used an estimator that creates weighted relationships between terms of distinct thesauri by pre-processing common instances from both gazetteers. To achieve this goal, we collect statistics about the intersection set of instances from the gazetteers to be integrated. Then, using the mapping rate estimation model, we reclassify all features migrated from one gazetteer to the other using the terms from the second thesaurus with the largest mapping rate estimation value.

Acknowledgements

This work is partially supported by CNPq under grants 550250/05-0, 140417/05-2 and 552068/02-0.

References

[1] ADL (1999), "Alexandria Digital Library Gazetteer", Map and Imagery Lab, Davidson Library, University of California, Santa Barbara. Available at: http://www.alexandria.ucsb.edu/gazetteer

[2] Breitman, K. K., Felicissimo, C. H. and Casanova, M. A. (2005), "CATO – A Lightweight Ontology Alignment Tool". Proc. 17th Conf. on Advanced Information Systems Engineering (CAISE'05), 2005, Porto, Portugal.

[3] Castano, S. et al. (2004), "Semantic Information Interoperability in Open Networked Systems". In: Proc. of the Int. Conference on Semantics of a Networked World (ICSNW), in cooperation with ACM SIGMOD 2004, Paris, France.

[4] Doan, A. et al. (2003), "Learning to match ontologies on the Semantic Web". In: The VLDB Journal - The International Journal on Very Large Data Bases, Volume 12, Issue 4, 2003. ISSN: 1066-8888. pp. 303-319.

[5] GNS (2006), "GEOnet Names Server", U.S. National Geospatial-Intelligence Agency, USA. Available at: http://gnswww.nga.mil/geonames/GNS.

[6] Hill, L. L., Frew, J. and Zheng, Q. (1999), "Geographic names: The implementation of a gazetteer in a georeferenced digital library." D-Lib (January 1999). http://www.dlib.org/dlib/january99/hill/01hill.html

[7] ISO-2788 (1986), "Documentation -- Guidelines for the development of monolingual thesauri", International Standard ISO-2788, Second edition -- 1986-11-15.

[8] Janée, G. and Hill, L. L. (2004), "ADL Gazetteer Protocol". Alexandria Digital Library Project. Retrieved Jul 28 2006. Available at http://www.alexandria.ucsb.edu/gazetteer/protocol/

[9] Leme, L.A.P.P. (2006) *Uma arquitetura de software para aplicações de catalogação automática de dados geográficos*. Dissertação de Mestrado. Departamento de Informática, PUC-Rio

[10] McGuinness, D. et al. (2000), "The Chimaera Ontology Environment". In Proceedings of the 17th National Conference on Artificial Intelligence (AAAI), 2000.

[11] Mena, E. et al. (1996), "OBSERVER: An Approach for Query Processing in Global Information Systems based on Interoperation across Pre-existing Ontologies". In: Proc. of the First IFCIS Int'l Conf. on Cooperative Information Systems, Brussels (Belgium), IEEE, pp. 14-25. Available at: http://sid.cps.unizar.es/PUBLICATIONS/POSTSCRIPTS/coopis96.ps.gz

[12] Mena, E. et al. (2000), "OBSERVER: An Approach for Query Processing in Global Information Systems based on Interoperation across Pre-existing On-

tologies", Int'l journal on Distributed And Parallel Databases (DAPD), 8(2):223-272, Kluwer Academic Publishers. Available at: http://sid.cps.unizar.es/PUBLICATIONS/POSTSCRIPTS/dapd00.ps.gz

[13] Noy, N. F. and Musen, M. A. (2003), "The PROMPT Suite: Interactive Tools For Ontology Merging And Mapping". International Journal of Human-Computer Studies, 2003.

[14] Percivall, G. (2003), OpenGIS® Reference Model, Document number OGC 03-040, Version 0.1.3, Open GIS Consortium, Inc.

[15] Spertus, E., Sahami, M. and Buyukkokten, O. (2005), "Evaluating Similarity Measures: A Large-Scale Study in the Orkut Social Network". In: Proceedings of the Eleventh ACM SIGKDD International Conference on Knowledge Discovery and Data Mining, Chicago, Illinois, USA, August 21-24, 2005. pp.678-684.

[16] UNESCO (1995), "UNESCO Thesaurus". United Nations Educational, Scientific and Cultural Organization, 1995. http://www.ulcc.ac.uk/unesco

[17] WordNet (2005), "WordNet - a lexical database for the English language". Cognitive Science Laboratory, Princeton University, Princeton, NJ – USA. Available at: http://wordnet.princeton.edu

WS-GIS: Towards a SOA-Based SDI Federation

Fábio Luiz Leite Jr., Cláudio de Souza Baptista, Patrício de Alencar Silva, Elvis Rodrigues da Silva

Departamento de Sistemas e Computação
Universidade Federal de Campina Grande

1 Introduction

The amount of spatial digital information that has been generated in the last few years is increasing rapidly due to many reasons. Firstly, in the field of Earth Observation Systems (EOS), huge projects from space agencies such as NASA and CEO are gathering petabytes of information per year in satellite images which have been accumulated with current stored data. Secondly, map agencies are producing their maps electronically. Lastly, advances in mobile devices and wireless infrastructure have motivated the large use of spatial related data.

There are several issues to be addressed in order to make these large datasets easily accessed. Firstly, these data should be indexed and retrieved by spatial-aware search engines, as current search engines use only text-based retrieval, which is unsuitable for spatial data [12]. Secondly, data replication is an important issue and should be minimized, as dealing with spatial data involves high costs not only in terms of acquisition, extraction, transformation and loading but also storage and maintenance. Finally, the use of standards is a mandatory issue, as they enable both data and services to be easily discovered and to interoperate. Hence, the implementation of a spatial data infrastructure – SDI – at global, national, and local scale, using largely adopted standards based on service-oriented architecture is a very hot research topic [10].

One of the first attempts of providing access to these spatial datasets was the design of clearinghouses. The main aim of a clearinghouse was to provide a centralized Web portal in which, by using metadata annotation on spatial data distributed over the Web, users may search, view and transfer such data [9]. Data providers should then subscribe into the clearinghouse in order to make their spatial data retrievable.

More recently, the advent of service-oriented architecture has motivated the adoption of geoportals, which consist of Web portals based on services, data, search engines and applications. These geoportals provide access to data and metadata and links to the service providers. Thus, there is a shift from a data-centric SDI to a service-oriented one [17].

Nowadays, the Web is used not only for document searching but also for the provision and use of services, known as Web services [3], which return dynamically changing data. Particularly in the spatial domain, examples of such services include but are not limited to basemaps, coordinate transformation, gazetteer, location-based systems, routing, overlay, buffer, and so on [10].

In the Web services domain these standards involve a common protocol for exchanging messages; a repository which enable to catalogue and search services; a description language for service annotation and a mechanism of service composition.

Several of such standards have been proposed. The SOAP (Simple Object Access Protocol) protocol specifies how Web services may encapsulate messages in XML documents. WSDL (Web Service Description Language) aims to describe service interfaces offered by Web services, and it may describe service operations, input and output parameters of each message, data type parameters and so on.

UDDI consists of a framework which enables the publication and discovery of services. Service providers use UDDI to announce their services; whereas clients may use it in order to find out a specific service, and how to interact with it. These standards are based on XML.

Currently there are several Web services published on the Internet, however, an important issue is how to find out precisely the service a user is requesting? Keyword-based search techniques implemented by search engines are not adequate for service discovery. Some reasons for this limitation include [3]:

- UDDI registry contains little textual information about the underlying services. Usually there is just a short description of the service provider;
- Keyword-based search may retrieve many irrelevant services just because they contain the addressed keyword. Also, a relevant service may not be retrieved just because this service uses synonyms or related terms to the used keywords.
- It is very complex to describe a given service based only on keywords;

- WSDL and UDDI descriptions express almost none of the service semantics;

Nonetheless, when dealing with spatial data these directories should become spatial-aware. For example, a user might be interested in a map with rivers inside a particular region.

The lack of semantics in SOA has been addressed recently [7, 20]. One possible solution is to enhance the catalogues with semantic through the use of ontologies, that represents a formal specification of a shared conceptualization [14]. The Open Geospatial Consortium (OGC) has proposed some standards in this direction, but they are very light concerning semantics. There are some proposals on the use of spatial domain ontologies to improve resource discovery on the Web [12].

In this paper we propose a service-oriented architecture for spatial data sharing. Hence, users may query a distributed catalogue service, semantically enhanced with ontologies. The remainder of the paper is organized as follows. Section 2 presents our architecture. Section 3 addresses some implementation issues through a real example. Section 4 highlights related work. Finally, section 5 concludes the paper and discusses further work that should be undertaken.

2 The WS-GIS Architecture

We propose WS-GIS which is a SDI based on service-oriented architecture. WS-GIS enables spatial search on distributed catalogues that form a federation of spatial databases. The architecture (see figure 1) enables access to spatial Web services and it has the following features:

- Resource location capability in federated catalogues: by using ontologies it is possible to model catalogue services, so that distributed searches, even spatial, on such catalogues may be executed. For instance, a fireman may pose the following query, when faced to a forest fire: "Which kind of vegetation and animal species might be affected by the fire spots identified in a particular satellite image?";
- Query caching: previous services chaining may be stored in the workflow for future reuse;
- Domain transparency: client should not care about where the resources come from. However, information about data provenance may be retrieved on demand.

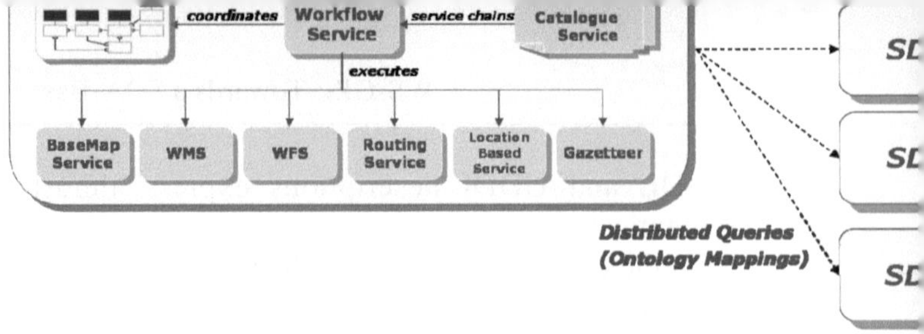

Fig. 1. WS-GIS overall architecture

In figure 1, catalogue service represents an OGC Catalogue which describes the underlying services using both services and ontologies. WFS and WMS represent the OGC Web Feature Serv Web Map Service, respectively [18]. LBS is a Location based ser context-aware applications [2], the routing service is responsible f finding and the Gazetteer service implements a gazetteer based Each SDI contains a set of spatial services provided by a given p private organization. The SDIs cooperate themselves through a ca federation. Hence, it is possible to obtain resources provided by an the federation from any client. In the proposed architecture, th agent might be able to query the client SDI catalogue; which may ute the query to other SDIs and it is responsible for composing th results and sending it to the client. In the following each of these are further detailed.

2.1 The Catalogue Service

The catalogue service is the core of our architecture, as it is respon SDIs integration at local, national or international scope. Each SD tains a spatial-aware catalogue which may find resources locally. T logue implements a communication protocol which propagates r queries to other SDI catalogues.

The catalogue communication protocol may choose one catalo vice to cumulate query results, which contain references to the s and send them to the workflow service. We use the following str maintain the federation:

- when a registry enters into the federation, the administrator mu into its registry an URL which already is part of the federation;

- after getting a registry address in the federation, the new registry sends a message to obtain the address table of the contacted site.

After that, all new addresses obtained by the new registry should be transmitted to the machines which are part of its address table. If there is no such new address, it will send an acknowledgment which states its permanency in the federation.

In order to comply with OGC standards, the catalogue service implements an OGC Catalogue API which contains getCapabilities, getRecords, getRecordById, describeRecord, getDomain, presentResults and transaction methods.

In the following the registry and query modules are presented, followed by a discussion on how these catalogues semantically interoperate.

2.1.1 The Registry Module

The catalogue maintains a metadata repository which describes the SDI underlying resources. These metadata are implemented using both service ontologies, which provide service semantics; and domain ontologies, which describe data semantics. After receiving a request, the catalogue queries the metadata repository in order to find the services which fulfill the request. The catalogue represents these ontologies using the Jena framework which stores them in a relational database system. The underlying data is stored in a spatial-aware database. This registry module uses the Jena framework to manipulate the ontologies.

2.1.2 The Query Module

The query module retrieves information based on query spatial features. Firstly, this module retrieves metadata that describe the available services and then, it performs spatial operations based on user requirements. Usually, the geographic information provided in the query is based on place names. Thus the catalogue may search for a Gazetteer service which may return the geographical coordinates of a given place name.

The query module also makes a query broadcast to the other catalogues in the federation. Hence, the catalogue service attempts to answer user query only regarding local catalogue data. In the case of partial fulfillment of user requirements, the catalogue propagates the query to other catalogue instances. The query propagation algorithm follows.

```
Algorithm search (query, hops) {
 if (localOntology != query.ontology)
  query = mapping (localOntology, query);
  // semantic matching among the SDI catalogues
 if (query is fully satisfied)
  return resultSet; // as service references
 else {
  resultSet = partialResult();
  newQuery = mapping (globalOntology, (query -
resultSet));
  if (hops threshold is not achieved) {
   resultSet = resultSet ∪ otherSDIs.search
(newQuery, hops);
   // If a timeout is achieved throws an excep-
tion
   return resultSet;
  } else
   // throws an exception;
 }
}
```

Code. 1. The query propagation algorithm

2.1.3 Semantic integration among catalogues

Ontologies are being applied very successfully in supporting information and knowledge exchange. However, for many reasons, different people and organizations will tend to use different ontologies. Unfortunately, the semantic heterogeneity also represents a drawback when different network catalogues make use of different domain ontologies. Hence, the system attempts to make ontology mappings between the SDI ontologies, trying to reconcile the concepts and properties related to them.

Each SDI defines its own ontology that is used in its catalogues. It means that there is a common vocabulary in each SDI that represents the concepts of the domain ontology used to describe the resources. However, we make an assumption that there is a set of concepts which can be shared among the SDIs, through clustering semantic communities. Thus, communities are formed through semantic mapping between SDIs. Each community has its own domain ontology which can be shared, and the other communities can cooperate in this scenario, mapping their domain ontologies. This can contribute to simplify the semantic mappings related to heterogeneity among SDIs [15].

Considering this scenario, the query module broadcasts both the rewritten query in the global SDI ontology. Thus, the catalogues that receive the propagated query make the semantic mapping between the local ontology and the global one. To accomplish this task it is necessary to verify in which relationship level these mappings are possible. Once the referred catalogues belong to the same community, we can make the assumption that they share the same ontology, which is the global one. In the context of this work, we make an assumption that all the communities reconcile their ontologies. Through the mapping files originated from the ontology reconciliations, the catalogues share the concepts used by the SDIs and interprets the network propagated queries [15].

2.2 The Workflow Service

The workflow service is also responsible for executing the selected service chaining. The catalogue service returns a record to the user that contains all the possible workflows that are able to fulfill the submitted request. The next step is the selection of the desired service workflow by the user. Then, the catalogue service activates the workflow service, providing the workflow selected by the client as parameter. The workflow service executes the selected workflow and accomplishes the monitoring phase that comprises basically aspects like checking the right functioning of the involved services, response time, service availability, unexpected behavior and error handling. Furthermore, for each service involved in the workflow, the service inputs and outputs are checked so that each output of a service is compatible with the input of the subsequent service in the workflow.

2.3 The LBS Web Service

Mobile users may register to receive alerts on subject of their interest while they are on the move. The LBS Web service is responsible for receiving and managing user context information. When a client is registered and provides profile, and appointments, it is able to receive context-aware information. The main available operations are:

- registerUser() which registers a user in a Web service, its profile and context information;
- updatePosition() which enables to update user geographic location in the system, and analyses the context searching for actions which are relevant to user context. Each time user location is updated the context is analyzed;

- registerAppointment() which registers a user appointment.

2.4 The Web Map Service and Web Feature Service

These services are based on the OpenGeospatial WMS and WFS. Nonetheless, we have implemented a Web service interface for such services. A request is done through several parameters in a URL. By implementing these services as Web services we not only make the requests easier, but also it is possible to publish these services in a directory services such as UDDI, in order to provide automatic service discovery and invocation.

Currently, we have implemented the following WMS operations: getCapabilities, getMap and getFeatureInfo. The getCapabilities method obtains service metadata which describes relevant information including maximum number of layers which may exist in a map, available layers and styles, coordinate reference systems (CRS), bounding box and scale. The getMap method returns a map according to the parameters received. These parameters include Layers, which contains the list of layers in a map; CRS, which contains the coordinate reference system; BBOX, which expresses the bounding box of interest; and Format, which contains the map output format. The getFeatureInfo method enables to retrieve more information about a chosen feature, for instance, a River name, length, quality of water, etc.

The Web Feature Service specification enables the user to query and update geospatial data in an interoperable way. This service contains the following operations: getCapabilities, describeFeatureType and getFeature. The getCapabilities method is responsible for describing available features, and the operations supported by each feature. The describeFeatureType method obtains the structure of a given feature type, which is described in an XMLSchema. Finally, the getFeature method enables users to specify which feature properties will be queried, as for example the name of a given feature.

2.5 The Routing Service

This service is responsible for providing on demand routes among two or more places. The client sends a request to this service (using the getRoute() method) providing two pairs of latitude and longitude coordinates. The client may also inform some preferences to be taken into account by the service such as road and traffic conditions.

The routing Web service stores data about routes and their intersections as a graph. The path between each two pairs is pre-computed and stored in the database, using the total materialization strategy. The storage cost using this approach is high, but as disk prices are decreasing, the gains obtained in processing power are worthy. We use a graph hierarchy to minimize the storage costs. The graph is partitioned into non-interleaving sub-graphs and there are nodes which take part in more than one sub-graph, they are called border nodes and they take part in the super-graph.

Paths are computed in each fragment and in the super-graph. We used the classical Dijkstra best path algorithm to pre-calculate all costs [11].

2.6 The Gazetteer Service

Gazetteer is used to help users to find bounding box coordinates from place names or feature types, and vice-versa. The simplest way of implementing a gazetteer is by having pairs of place names and spatial footprints. In this case, the gazetteer does not consider the feature type of each element. The feature types are either implicit or explicit. In the former case, it is assumed that the user is familiar with the place name type, for example, England is a country. In the latter, the feature type comes together with the place name, as, for example, North Sea.

A reasonable core gazetteer should include feature type together with place name and footprint. This enables determining, for example, whether bank is a building or a land formation along the edge of a river [19].

We have implemented a Gazetteer Web service which contains place names, spatial footprints, temporal interval, feature types, and lineage. The first two attributes are mandatory, whilst the latter three are optional. Place name contains the name of the place being identified; spatial footprint contains a bounding box represented by two pairs of latitude/longitude points; time is an interval that contains two timestamps: the beginTime and endTime of that place name; feature type contains a feature that characterizes the place name; and lineage contains the source that provided that information.

This service is accessed via searching one of the three attributes: place name, space and feature type. For example, a user may be interested in where a determined place is, what exists in a determined spatial footprint, or where a particular feature type exists on the Earth's surface.

2.7 Implementation issues

The Web services were implemented using the Java J2EE and JAX-RPC. Each service was developed in three steps: interface definition, which contains the methods which may be called by a client; interface implementation, which contains the service logic; and the service deployment in an application server. We have used the Java Sun Application Server. After the second step is done, the WSDL file is generated automatically.

All the implemented data services (WMS, WFS) and the processing services (LBS, routing services) use PostgreSQL with Postgis DBMS. The catalog service also uses PostgreSQL DBMS to materialize the ontologies and the Jena framework as inference engine to verify the ontology consistence.

3 An Example Scenario for the WS-GIS

In the Brazilian State of Paraíba there are many state departments and agencies which deal with data on natural resources and infrastructure. They manipulate information about water resources, weather, roads, etc. Unfortunately the underlying information systems do not interoperate and there is too much replication. Obviously this replication introduces inconsistency and results in a waste of space on disk, especially when dealing with spatial data such as satellite images and vector maps which require huge space on disks. Let us suppose that there exist three departments, each one with its SDI:

- the Water Management Agency (AESA), which deals with rivers, dams, lakes, poles, and so on;
- the Environmental Management Agency (SUDEMA), which deals with quality of water (rivers, dams, lakes, poles and so on); and
- the infrastructure agency (DER) which deals with road maintenance.

In these SDIs there is data replication: bodies of water are replicated in SUDEMA.

Suppose the a user would like to visualize a SVG (a W3C Specification based on XML to display graphics on a browser) map which contains the regions of Paraíba, the rivers (and its water quality) in Sertão region, and roads of the entire State, even whether these data layers are distributed in many agencies. Obviously, transparency on distribution is a mandatory issue [13]. Moreover, let us suppose that the AESA SDI contains a Gazetteer service and a MapConvertion service which receives a GML map and returns it in SVG.

The user interaction follows. Firstly, the client executes, for example in the AESA SDI, the getRecords method, with OGC Filter parameters (Code 2), in the catalogue service requesting the desired information (e.g. a map with regions of Paraiba, rivers and information on water quality in Sertão region and roads). In order to pose the query the user uses concepts defined in the ontology which pertains to the local SDI. After receiving the query, the AESA catalogue service proceeds as follows, according to the algorithm presented in section 2:

1. it queries the Gazetteer service of the local SDI, in order to obtain the State region geometries to query the service registries using the intersection with rivers and roads;
2. it verifies in the local SDI if it contains services with the requested data. In order to do that, the catalogue runs a spatial query searching for the rivers in the Sertão region. water quality data and roads;
3. as the AESA SDI does not have the whole requested data set, the query is propagated to other SDI catalogues. However, the local catalogue rewrites the original query in order to remove from it the data which is already found in its database. Then, the catalogue rewrites the query searching for roads and quality of water data in the geographic area of interest and propagates this query to other catalogues.

```
<ogc:Filter>
 <ogc:Or>
  <ogc:And>
   <PropertyIsEqualTo>
    <PropertyName>FeatureName</PropertyName>
    <Literal>River</Literal>
   </PropertyIsEqualTo>
   <PropertyIsEqualTo>
    <PropertyName>hasQuality</PropertyName>
    <Literal>yes</Literal>
   </PropertyIsEqualTo>
   <Overlaps>
     <PropertyName>Geometry</PropertyName>
     <gml:Polygon
srsName="urn:ogc:def:crs:EPSG:6.6:4618">
      <gml:exterior>
       <gml:LinearRing>
        <gml:posList>
```

```
      -37.001821,-7.175744  -36.992499,-7.180127  -
36.976542,-7.160474   ...   -37.019035,-7.190465  -
37.001821,-7.175744
           </gml:posList>
          </gml:LinearRing>
        </gml:exterior>
       </gml:Polygon>
     </Overlaps> </ogc:And>
    <ogc:And>
     <PropertyIsEqualTo>
      <PropertyName>FeatureName</PropertyName>
      <Literal>Road</Literal>
     <PropertyIsEqualTo>
      <gml:posList>
       -38.76516,-6.91093    -38.75917,-6.9155    -
38.75912,-6.91657     ...    -38.75622,-6.9036    -
38.76516,-6.91093
      </gml:posList>
    </ogc:And>
   </ogc:Or>
  </ogc:Filter>
```

Code. 2. The catalogue service query using OGC Filter

It is important to notice that, when the SUDEMA catalogue receives the query, it needs to perform the ontology matching of the concepts which come from AESA with its own ontology concepts, as they probably use different ontologies. Hence, there is a State ontology which has mappings for each agency ontology. Figure 2a presents part of AESA ontology which is the same of the State one. Figure 2b presents the SUDEMA ontology. It is important to notice that in the AESA ontology, and then in the State one, the body of water class is superclass of dam, river and pole classes. Nonetheless, in the SUDEMA ontology there is no river, dam and pole concepts, all of them are treated in a unique class called body of water.

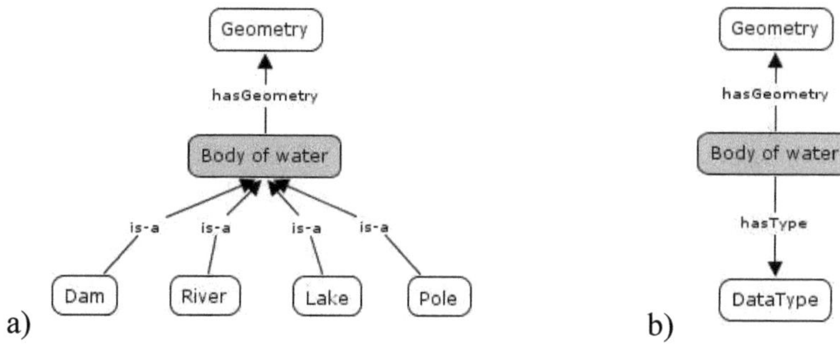

Fig. 2. a) AESA and State ontology; b) SUDEMA ontology

In the following there is an example of ontology mapping between SUDEMA and State using OWL. Thus, the concept body of water in SUDEMA is mapped directly to the one in the state ontology. Hence, the "river" concept in AESA ontology (and in the state) is mapped to the "body of water" concept in the SUDEMA ontology through "hasType" property, defined as "river" DataType.

```
<owl:Class rdf:ID="BodyOfWater1SUDEMA">
<owl:equivalentClass rdf:resource="BodyOfWaterAESA"/>
</owl:Class>
```

Code. 3. Ontology mapping between SUDEMA and State using OWL

The following OWL code indicates that the SUDEMA BodyOfWater class is the union of lake, river, pole and dam in the State ontology, and consequently in the AESA one.

```
  <rdf:RDF
xmlns="http://localhost/ontologias/sudema.owl#"
   xml:base="http://localhost/ontologias/sudema.owl"
   xmlns:globalSDI="http://localhost/ontologias/
globalSDI.owl#"
   xmlns:xsd="http://www.w3.org/2001/XMLSchema#"
   xmlns:dc="http://purl.org/dc/elements/1.1/"
   xmlns:rdfs="http://www.w3.org/2000/01/rdf-schema#"
   xmlns:daml="http://www.daml.org/2001/03/daml+oil#"
   xmlns:rdf="http://www.w3.org/1999/02/22-rdf-syntax-
ns#"
   xmlns:owl="http://www.w3.org/2002/07/owl#">
   <owl:Ontology rdf:about="">
    <owl:imports
rdf:resource="http://localhost/ontologias/aesa.owl"/>
   </owl:Ontology>
```

```
<owl:Class rdf:ID="Body_of_Water">
 <rdfs:subClassOf rdf:resource="&owl;Thing"/>
 <rdfs:subClassOf>
  <owl:Class>
   <owl:unionOf rdf:parseType="Collection">
    <owl:Class rdf:about="&globalSDI;Dam"/>
    <owl:Class rdf:about="&globalSDI;Lake"/>
    <owl:Class rdf:about="&globalSDI;Pole"/>
    <owl:Class rdf:about="&globalSDI;River"/>
   </owl:unionOf>
  </owl:Class>
 </rdfs:subClassOf>
</owl:Class>
<owl:DatatypeProperty rdf:ID="hasType"/>
<owl:Class rdf:ID="Geometry"/>
</rdf:RDF>
```

Code. 4. Ontology mapping between the BodyOfWater of SUDEMA and the State one using OWL

In parallel, the DER catalogue also receives the request from AESA, and it solves the semantic differences between concepts and verifies in its metadata repository that it contains the Paraíba roads. Thus, it rewrites the query using the State ontology and broadcasts it to other SDI (in this case, the query concerns only quality of water). Lastly, it returns the references of the data services to the AESA catalogue.

After receiving all services references, the query module mounts a service chaining and sends it to the client. This workflow is presented in figure 3.

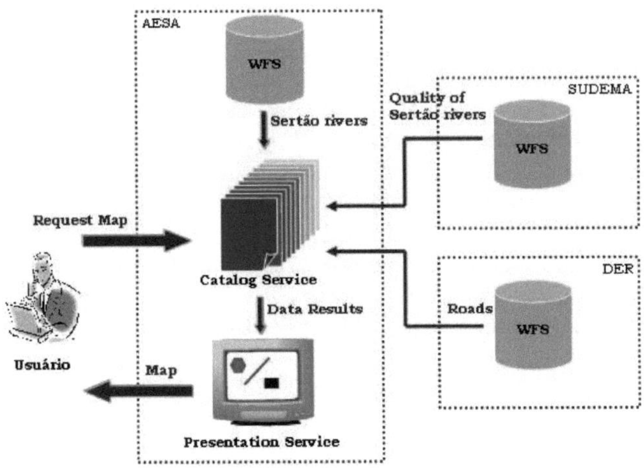

Fig. 3. Service chaining of the query

The workflow service executes the service chaining and monitors the behavior of the involved services. Also, it sends the inputs and outputs necessary to each service. The final result is returned to the client which may render the map as it is presented in figure 4, using the framework iGIS [4].

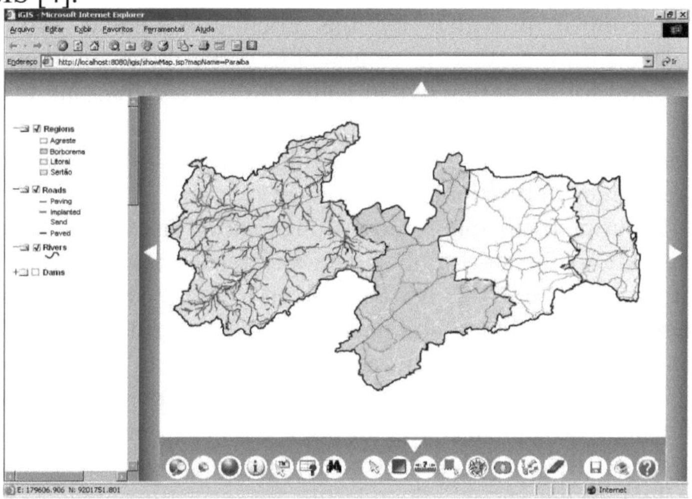

Fig. 4. Query result on a map tool

4 Related Work

Bernard and Craglia [5] present a survey on SDI and discuss its evolution from clearinghouses to geoportals. They emphasize the importance of migrating to a service-oriented infrastructure, in which Web services play a central role. Moreover, efforts on standards for these services are highlighted. These efforts result in a (GI)-Service Driven Infrastructure which characterizes the next generation of SDI which is focused on distributed and shared functionalities. Also, new functionalities can be achieved by service chaining.

The provision of service chaining enables scalability without compromising performance. There are several issues which should have taken into account such as the level of transparency of the chaining to the client and the effort required in the client to implement service coordination. According to Alameh [1], there are three types of service chaining:

- Client-Coordinated service chaining: the client defines and controls the service execution order in the chaining;

- Static Chaining using aggregate services: the service chaining is hidden from the client, so the latter has only a link to the chaining. However the chaining is not dynamic;
- Workflow-managed service chaining with mediating services: combines the simplicity of static chaining using aggregate services with the flexibility and control provided by the client-coordinated service chaining. The workflow service is responsible for error manipulation and exceptions, and for mounting dynamically the service chaining to be used by clients.

Although, Alameh deals with service chaining, there is no concern on locating spatial recources in distributed catalogues.

Zhao et al. [21] propose an integration model which uses a Metadata CatalogueService (MCS) for storing metadata and retrieving resources in grids; and the model OGC ebRIM, which enables to aggregate semantics to the spatial queries in distributed architecture based on grids. Their idea is to implement semantic mapping between these two models by using ontologies in OWL. Therefore, user poses queries at the semantic level and those queries are mapped into the grid through MCS. Nonetheless, this work does not address the mapping between OGC Catalogue Services – ebRIM and the UDDI registries, which underpin the service-oriented SDIs.

Sivashanmugam et al. [20] propose a peer-to-peer network of public and semi-private UDDI registries, which enables transparent access to any registry in the federation. They also use ontologies to classify the registries and to locate resources. This work does not deal with spatial modeling of OGC catalogues, so that spatial queries are not fully accomplished.

Boucelma et al. [6] integrate data and query language on heterogeneous spatial data sources distributed in a unique view. Their proposal is based on mediators and wrappers which access distributed data and implement query mapping. They use the global-as-view model [8], in which the wrappers enable not only answering queries on distributed data but also querying local data. The wrappers are built using components which interact with Web Feature Service. However, this system is based on a centralized architecture. Also, the inclusion of new services must be done manually.

Hübner et al. [16] propose to unify distributed data sources in a unique point of interaction with the client, through a map definition tool. Their system uses metadata for resource location, which is classified using ontologies. Hence, semantic and syntactical searches may be achieved even with spatial data. However, the proposed architecture demands great effort from users. Moreover, data locating service is not fully compatible with OGC which penalizes interoperability. Also, the approach maintains a centered catalogue.

The main contribution of our work is concerning the design of a SOA-based SDI which aims to integrate, locate, and catalogue distributed spatial data sources (services and data). Mainly, we propose a catalogue federation which enables dynamic exchange of heterogeneous SDIs. The only requirement is to follow OGC recommendation. Hence, users have an integrated and transparent view of the available resources which makes easy data exchange and minimizes data replication. Moreover, by using Web services we provide interoperability, portability, flexibility and dynamic chaining, which are not present in geoportals and clearinghouses. Finally, the WS-GIS architecture combines GI Service composition with a service catalogue federation, based on the ontologies, which enables a semantic modeling of catalogues.

Furthermore, the proposed communication protocol is spatial-aware. Hence, it is possible to rewrite queries on the fly so that only the reminding data is forwarded to other catalogues in the federation. This spatial query optimization reduces data redundancy in the result set. To the best of our knowledge this approach has not been proposed yet in the specialized literature.

5 Conclusion

The large availability of spatial datasets and the advent of service-oriented architectures have motivated research on SDI based on Web services. In this paper we presented the WS-GIS architecture which aims to implement a SOA-based SDI which promotes a set of spatial services and a federation of catalogues to enable distributed queries. The main purpose of such architecture is to enable integration of heterogeneous spatial data sets, by promoting interoperability, flexibility, service chaining and scalability. We have also built a prototype to work in a Local SDI, in a real scenario of managing water resources in the Brazilian state of Paraíba.

As further work we intend to improve the workflow service so that it can take into account load balancing, quality of service, data provenance and service context. Moreover, we intend to implement other spatial services such as OGC WCS; to migrate our Gazetteer Web service to be OGC compliant; and to analyze the use of grid computing in our architecture.

References

[1] Alameh N (2003) Chaining Geographic Information Web Services. In: IEEE Internet Computing 7 (5): 22-29

[2] Almeida DR, Baptista CS, Silva ER, Campelo CEC, Figueiredo HF, Lacerda YA (2006) A Context-Aware System Based on Service-Oriented Architecture. In: 20th International Conference on Advanced Information Networking and Applications. IEEE Computer Society, pp 205-210

[3] Alonso G, Casati F, Kuno H, Machiraju V (2004) Web Services: Concepts, Architectures and Applications. Springer-Verlag

[4] Baptista CS, Silva ER, Leite Jr. FL, Paiva AC (2004) Using Open Source GIS in e-Government Applications. In International Conference on Eletronic Government. Lecture Notes in Computer Science, pp 418-421

[5] Bernard L, Craglia M (2005) SDI - From Spatial Data Infrastructure to Service Driven Infrastructure. In: First Research Workshop on Cross-learning on Spatial Data Infrastructures and Information Infrastructures. Enschede, Netherlands

[6] Boucelma O, Essid M, Lacroix Z (2002) A WFS-based mediation system for GIS interoperability. In: 10th ACM International Symposium on Advances in Geographic Information Systems. ACM Press, pp 23-28

[7] Burstein M, Bussler C, Zaremba M, Finin T, Huhns MN, Paolucci M, Sheth AP, Williams S (2005) Semantic web services architecture. In: IEEE Internet Computing 9 (5): 72-81

[8] Chawathe S, Garcia-Molina H, Hammer J, Ireland K, Papakonstantinou Y, Ullman JD, Widom J (1994) The TSIMMIS Project: Integration of heterogeneous information sources. In: 16th Meeting of the Information Processing Society of Japan, pp 7-18

[9] Crompvoets J, Bregt A, Rajabifard A, Williamson I (2004) Assessing the worldwide developments of national spatial data clearinghouses. In: International Journal of Geographical Information Science 18 (7): 665-689

[10] Davis C, Alves LL (2005) Local Spatial Data Infrastructures Based on a Service-Oriented Architecture. In: Brazilian Symposium on GeoInformatics. SBC.

[11] Dijkstra EW (1959) A note on two problems in connection with graphs. In: Numerische Mathematik 1: 69-271

[12] Fu G, Jones CB, Abdelmoty AI (2005) Building a Geographical Ontology for Intelligent Spatial Search on the Web. In: Databases and Applications, pp 167-172

[13] Garcia-Molina H, Ullman JD, Widom JD (2001) Database Systems: The Complete Book. Prentice Hall

[14] Gómez-Perez A, Fernández-López M, Corcho O (2004) Ontological Engineering: with example from the areas of Knowledge Management, e-commerce and the Semantic Web. Springer-Verlag

[15] Hameed A, Preece A, Sleeman D (2004) Ontology reconciliations. Staab S, Studer R (eds) Handbook on Ontologies. Springer-Verlag, Germany

[16] Hübner S, Spittel R, Visser U, Vögele TJ (2004) Ontology-Based Search for Interactive Digital Maps. In: IEEE Intelligent Systems 19 (3): 80-86
[17] Maguire DJ, Longley PA (2005) The emergence of geoportals and their role in spatial data infrastructures. In: Computers, Environment and Urban Systems 29 (1): 3-14
[18] Open Geospatial Consortium (2006). http://www.opengeospatial.org
[19] Pazinatto E, Baptista CS, Miranda RAV (2002) GeoLocalizador: um Sistema de Referência Espaço-Temporal Indireta utilizando um SGBD Objeto-Relacional. In: Brazilian Symposium on GeoInformatics. SBC.
[20] Sivashanmugam L, Verma K, Sheth A (2004) Discovery of Web Services in a Federated Registry Environment. In: IEEE International Conference on Web Services. IEEE Computer Society, pp 270+
[21] Zhao P, Chen A, Liu Y, Di L, Yang W, Li P (2004) Grid metadata catalogue service-based OGC Web registry service. In: 12th Annual ACM International Workshop on Geographic Information Systems. ACM Press, pp 22-30

Electricity Consumption as a Predictor of Household Income: a Spatial Statistics Approach

Eduardo de Rezende Francisco[1,2], Francisco Aranha[2], Felipe Zambaldi[2], Rafael Goldszmidt[2]

[1]AES Eletropaulo
[2]Escola de Administração de Empresas de São Paulo
Fundação Getulio Vargas

This study is one of a series of research analyses carried out within the Integrated Research Program: Microcredit for Low Income Families in the city of São Paulo, supported by GVpesquisa (Getulio Vargas Research), a research agency maintained by Escola de Administração de Empresas de São Paulo da Fundação Getulio Vargas (FGV-EAESP). The authors would like to thank GVpesquisa for funding their research.

This chapter investigates the relations between electricity consumption, economic classification and household income by comparing the Brazilian Census Micro-Data with the customer database of AES Eletropaulo, one of the largest Brazilian electricity distribution companies, in the City of São Paulo. The research methodology was based on classical statistics and spatial auto-regressive models. Income and economic classification are recognized as efficient proxies for purchasing power. Income indicators based on electricity consumption, which may be generated by electricity companies using Geographic Information Systems (GIS), are found to be accurate in predicting income under a spatial-statistics approach. These findings also reveal a potential business development for electricity companies based on such indicators.

1 Introduction

Income is usually adopted as the main variable in studies of poverty and living conditions, since it provides access to basic goods and services [5]. It can be defined as the total earnings provided by work and other sources [16]; this concept may be applied to individuals, families and households. However, it is difficult to collect data about income; its declaration is frequently inaccurate and its value is subject to seasonal changes, becoming thus an unstable indicator for market researches [5].

Research professionals tend to prefer capturing indicators about economic classification and purchasing power based on possessions and educational levels as proxies for income and welfare. One conspicuous example of such an indicator is the Brazilian Economic Classification Criterion (CCEB – Critério de Classificação Econômica Brasil), or simply the Brazilian Criterion, created in 1996 by the National Research Enterprises Association (ANEP – Associação Nacional de Empresas de Pesquisa). The Brazilian Criterion, however, needs to be interpreted according to regional contexts [1], and is recognized as inadequate for characterizing families lying on the extremes of the income distribution [20, 23].

There is also criticism regarding the Brazilian Economic Classification Criterion due to operational difficulties. First, the information provided by the decennial census carried out in Brazil needs to be updated; and second, the Criterion needs adjustments when it is applied to specific regions or social segments.

We postulate that CCEB can be made more accurate by the inclusion in its formation of variables that bring additional information on purchase power. One of those variables is the consumption of electric energy, a utility used by 97% of the Brazilian households, a share that increases to 99.6% in urban areas [16]. Electricity is supplied to more households in Brazil than telecommunications, water and gas services [16]. Databases of electricity distributors usually contain consumption information about all their customers [8]. When aggregated into geographic, historic and seasonal datasets, electricity consumption indicators may contribute to the classification of consumers, even in regions where it is hard to collect data about income. Better consumer classification allows clearer target segmentation. Therefore, finding relations between electricity consumption and income may serve the interests of marketing and research professionals [12].

This chapter explores the relationships among three constructs: household income; economic classification based on the Brazilian Criterion; and electricity consumption. The purpose of this study is to investigate the convenience of using electricity consumption to socially characterize families in the city of São Paulo. It is also to verify if electricity consumption may serve to refine the Brazilian Criterion, so that market agents can clearly identify and attend their target segments. Since there are not many studies directly related to this objective, this is a seminal research which will hopefully foster future investigation.

The constructs and indicators of interest are described and examined in this chapter, and so are the data sets and variables used in the research. An analysis of the relations among the constructs is supported by regression modeling and the spatial structure of the constructs is explored by spatial auto-correlation techniques. As a result, a spatial auto-regressive model is developed, handling non-normality and heteroscedasticity of residuals. The main conclusion is that electricity may be an efficient predictor of income, but in the studied context it correlates better with high values of income than with lower ones; economic classification, on the other hand, always maintains its high correlation with income. Potential applications of the results are discussed at the end of the chapter.

2 Objective

The main purpose of this chapter is to investigate the relations between electricity consumption and family income; and between income and economic classification. The following three hypotheses are postulated:

- H_1: The higher the score of the Brazilian Criterion (Economic Classification), the higher the family income in the city of São Paulo;
- H_2: The higher the consumption of electric energy, the higher the family income in the city of São Paulo.
- H_3: There is a spatial dependence pattern of Household Income in the city of São Paulo, represented by decreasing levels of income from downtown to the suburbs.

The studied object is the population of the city of São Paulo, and the objective is to characterize the postulated relations in a territorial aggregate level, according to the weighted areas defined on the Demographic Census of 2000 by the Brazilian Institute of Geography and Statistics (IBGE – Instituto Brasileiro de Geografia e Estatística). The Micro-Data of the Demographic Census of 2000 is jointly used with information of electricity consumption for August 2000, in the city of São Paulo. The population of the city of São Paulo totaled 10,435,546 inhabitants in 2000, corresponding to 3,131,389 families [14]. A preliminary analysis of the data, including income, social exclusion and education indicators highlights a pattern of socioeconomic vulnerability from downtown to the suburbs.

3 Definitions

The main concepts used in this study are presented below; the relations between the constructs are described throughout the following sections.

3.1 Family, Household and Income

The concept of family used in this study is the same as defined by IBGE [17]: "groups of relatives or people living in the same household unit, depending on each other by means of domestic or living rules; or one person living alone in a household unit". Household is defined as a separate and independent "dwelling, constituted of one or more rooms" [17]. For reasons of parsimony, a correspondence between household and family is assumed, based on the very rare occurrence of more than one family living in a household [15].

Income is defined as the summation of all earnings provided by work and other sources [16] and may be calculated for individuals, families or households. It includes gross income (before taxation) from work, pensions, government and public social security programs (such as revenue programs, school grants, or unemployment benefits); and also rent of any type. It is measured in reais (R$, the Brazilian currency). However, collecting income data is often difficult and inaccurate.

3.2 Economic Classification and the Brazilian Criterion

In 1996, the Brazilian Criterion of Economic Classification (CCEB), or simply the Brazilian Criterion, was created by ANEP. It is based on indicators of ownership of goods and on indicators of head of the family educational level, composing a scale varying between 0 and 34 points, which are used to segment families into seven economic classes, named from the top to the bottom as A1, A2, B1, B2, C, D and E [1].

In 2004, the Brazilian Research Enterprises Association (ABEP – Associação Brasileira de Empresas de Pesquisa) was founded and became officially responsible for the norms of the Brazilian Criterion standards [1]. The Criterion is frequently criticized because of its limitation in segmenting populations according to lifestyle or social classifications [23]. The population distribution of the Brazilian Criterion score in metropolitan areas shows expressive variability among different regions in Brazil, reflecting therefore its capacity to discriminate purchasing power among different geographic locations [1].

3.3 Electricity Consumption

Electricity is the flow of electrical power or charge. It is a secondary energy source which means that we get it from the conversion of other sources of energy, like coal, natural gas, oil, nuclear power and other natural sources, which are called primary sources. In electrical engineering, power consumption refers to the electrical energy over time that must be supplied to an electrical appliance to maintain its operation [24].

Electricity corresponds to 64.2% of the energy consumption in Brazilian households, and this share shows a growth trend [2] caused by the increasing use of technology in households in the Southeast region, particularly computers, televisions and side-by-side refrigerators [18]. Studies show that the variability of electricity consumption among households is related to income levels [19, 21, 4]. Pompermayer and Charnet [21] found that social and demographic factors are also statistically significant influences.

Of the 5 million households in the AES Eletropaulo customer base, approximately 9.7% are classified as low-income consumers; and 2.4% are classified as illegal consumers (users of fraudulent connections) located on subnormal gatherings (slum areas) [6]. Together, these two groups correspond to 625,000 poor families serviced by the company [10].

4 Data Collection and Operational Aspects

Next are presented the operational aspects of the variables used in the study; then the investigated universe is characterized.

4.1 Micro-Data of the Demographic Census 2000

Official statistics are fundamental in a democracy, providing the Government and the society with data about the economic, demographic, social and environmental situation of the population. The Brazilian Institute of Geography and Statistics (IBGE) is among the main sources of official data in Brazil. The Brazilian Demographic Census occurs every ten years, maintaining a retrospective of the population characteristics since 1890. Its information is confidential and may be used only for research purposes.

The Census data collection was carried out between August and November 2000, in 54,265,618 households of the 5,507 municipalities of the Brazilian territory, divided into 215,811 areas [15]. Each area is named a census sector and covers from 200 to 300 households.

There are two researches included in the Demographic Census of 2000: (i) the Universe Research, which captures the characteristics of the households of the whole population (census); and (ii) the Sample Research, more complex and applied to about 11.7% of the private households in Brazil; it contains detailed information about the households and its residents such as education, religion, work activities and income.

The Sample Research cannot provide statistical significance at the level of census sectors and therefore its information is aggregated into weighted areas, which are mutually exclusive groups, and to which procedures of balanced estimation are applied in order to make inferences for the whole population [14]. The maximum size of a weighted area is a municipality and its minimum size is of 400 households, even if that implies noncontiguous areas, always respecting homogeneity characteristics such as income, number of people living in a household, infra-structure and educational level of the heads of the households. Only 482 municipalities had more than one weighted area; the other 5,032 constitute an area each one.

There are 9,336 weighted areas in Brazil [14]; their information refers to July 31st 2000. The city of São Paulo was divided into 96 districts, which are territorial and administrative units under the same judicial and fiscal administration [22] and represent the basis for the creation of the weighted areas with the support of geo-referenced computational systems. The city was divided into 13,278 census sectors and into 456 weighted areas [14].

Information about the Sample Research is provided by IBGE through the Micro-Data of the Demographic Census of 2000. Each reference unit is a household and allows for aggregation on the levels of weighted areas and districts. There are two separate micro-data files: (i) the household data set; and (ii) the people data set. Interviews were made in 30,669 households in the city of São Paulo, representing the universe of 3,032,905 units. Information of about 1,057,086 people was provided, representing the universe of 10,414,207 people.

The operational relation between the two data sets (households and people) was determined by a household code. The collected variables don't allow any association with census sectors, but they permit association with the weighted areas. Table 1 describes the main variables in the micro-data databases.

Table 1. Main variables of the Micro-Data of the Demographic Census of 2000

Data File	Variables
Household	Weighted Area Code; Census Sector Situation; Household Situation; Household Code; Number of residents; Number of Bathrooms; Toilette Vase, Electric Illumination, Radio, Refrigerator or Freezer, Videocassette player, Washing Machine, Microwave Oven; Telephonic Line; Micro-Computer; Number of Televisions; Number of Private Automobiles; Number of Air-Conditioners; Monthly Income of the Household
People	Household Code; Ordinal Number of Researched Person; Weighted Area Code; Household Situation; Relation between the Interviewee and the Head of the Household; If one knows how to read and write; If one goes to school or day-nursery; Current Course; Current Grade; Highest level Attended in a Course, having been approved at least in one grade; Highest Grade Approved; If one has concluded the attended course; Highest Level of Concluded Course Code; Years of Study; If one lives with a partner; Marital Status; If one hasn't any income provided by work; Total income provided by one's main job; Is one hasn't any income provided by other jobs; Total income provided by other jobs; Total income provided by work; Income provided by retirement or pensions, rentals, family pension or donations, public programs, and other sources; Total income

4.2 Income and the Adjusted Brazilian Criterion

In this study, the following representatives of income were computed: (i) total family income provided by main job; (ii) total family income provided by work; (iii) total family income (summation of all types of family income); (iv) total household income provided by main job; (v) total household income provided by work; and (vi) total household income (summation of all types of income of the household). Total household income is the summation of the income provided by all the residents of the family, even those who aren't considered family by IBGE.

Data of the People Database provided information about the educational level of the head of the family, part of the definition of Brazilian Criterion.

The data of the Sample Research did not include all the information necessary to calculate the exact Brazilian Criterion, as shown in Table 2.

Adjustments had to be made, resulting in a scale varying between 0 and 29 – the Adjusted Brazilian Criterion.

Table 2. Comparison of information used in the calculation of the Brazilian Criterion with the information available at the Sampled Micro-Data of the Demographic Census 2000

Brazilian Criterion	Sample Micro-Data
Number of color television sets	Number of television sets
Number of Radios	Radio Ownership
Number of Bathrooms	Number of Bathrooms and Ownership of toilette vase
Number of automobiles	Number of automobiles for private/personal use
Number of Monthly Employees	Number of individuals whose relation with the head of the family is classified as a "Domestic Employee"
Number of vacuum cleaners	Not asked
Number of Washing Machines	Ownership of Washing Machine
Number of Videocassette or DVD players	Ownership of Videocassette player
Number of Refrigerators or Freezers	Ownership of Refrigerator or Freezer

4.3 Electricity Consumption Data from AES Eletropaulo

The electric energy distribution companies maintain geo-referenced registers of its customers, including data such as consumption history and payment information. Each residential customer corresponds practically to a household. AES Eletropaulo, the main electricity provider in the City of São Paulo, controls its registers by a customer code which never changes, even though the customer may move to another household [9]. The data in this paper was collected from AES Eletropaulo databases between September 1999 and August 2000, and it was associated to the Micro-Data of the Sample Research of the Demographic Census of 2000. Two variables were then computed: (i) consumption during August 2000 and (ii) monthly consumption (mean) during the investigated period.

IBGE provides the description of the weighted areas of the Demographic Census of 2000. Using spatial algorithms [7], it was possible for

AES Eletropaulo to associate its customer codes to a weighted area and therefore compute the number of active customers in each area. All forms of computed income were then jointly analyzed with the Adjusted Brazilian Criterion and the two computed indicators of electricity consumption, for each weighted area in the city of São Paulo. Analyses were performed on the arithmetic means of the constructs in each weighted area.

5 Results and Analysis

The residential customers of AES Eletropaulo totaled 3,037,992 in August 2000; the total households pointed out by the Demographic Census of 2000 were 3.039.104, which suggests that the association was proper. Figure 1 and Table 4 describe the main statistical information of the three constructs.

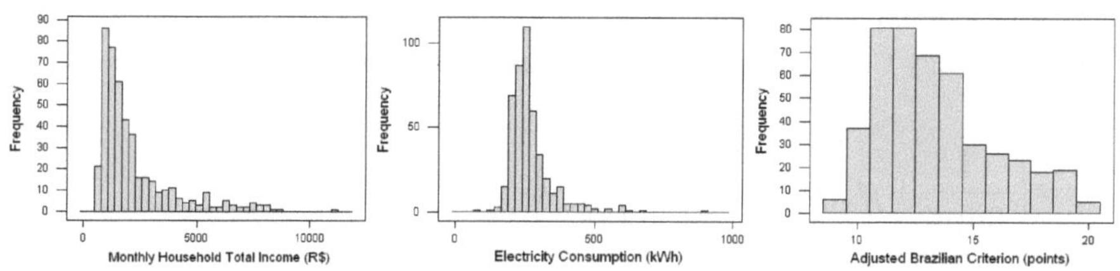

Fig. 1. Histograms of the three constructs of interest: Household Income, Electricity Consumption, and the Adjusted Brazilian Criterion

Table 4. Descriptive Statistics of: Household Income, Electricity Consumption, and the Adjusted Brazilian Criterion

Variable	Minimum	Maximum	Mean	Standard Deviation
Monthly Household Income (mean) (R$)	53,73	11.196,85	2.189,34	1.713,34
Electricity Consumption (mean) (kWh)	64,3	809,2	238,2	79,7
Adjusted Brazilian Criterion (mean) (points)	8,515	19,755	13,347	2,519

Because this is a seminal study, previous models weren't available and, therefore, many regression models were performed to test the hypothesis H1 and H2, maintaining income as a dependent variable and the Adjusted Brazilian Criterion and electricity consumption as independent variables. The two following models produced the best fits.

$n = 456$

Regression 1: Adjusted Brazilian Criterion as an Income predictor

y : Household Income (R$)

x : Adjusted Brazilian Criterion (points)

$$\hat{y} = \beta_0 + \beta_1 x + \beta_2 x^2 = 7512.63 - 1357.36\ x + 69.30\ x^2 \tag{1}$$

$R^2 = 0.960$; Adjusted $R^2 = 0.960$

Regression 2: Electric Energy Consumption as an income predictor

y : Household Income (R$)

x : Electricity Consumption (kWh) Obs: 8,600 is the maximum of the curve

$$\hat{y} = \left(\frac{1}{8600} + \beta_0\ \beta_1^{\ x}\right)^{-1} = \left[\frac{1}{8600} + 0.01412\ (0.98665)^x\right]^{-1} \tag{2}$$

$R^2 = 0.910$; Adjusted $R^2 = 0.853$

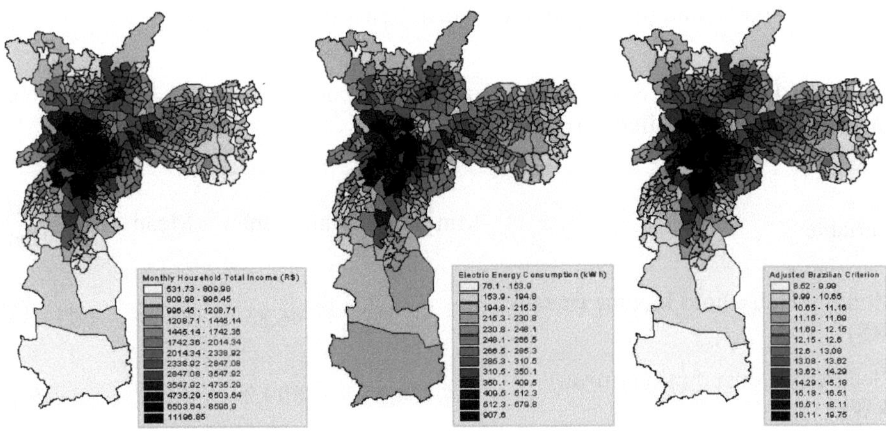

Fig. 2. Maps of the City of São Paulo representing: (i) Monthly Household Total Income, (ii) Electricity Consumption, and (iii) Adjusted Brazilian Criteria, per Weighted Areas

Both models are adequate and present R2 coefficients higher than 90% (corroborated by maps in Figure 2), which represent the percentage of data variability explained in the models. Both regressions are significant and so are their coefficients, what is confirmed by the F statistics and P-values of each model. The residual errors, on the other hand, show that their magnitudes tend to increase as the income levels do; and so their structure is probably more complex than the models describe. The Kolmogorov-Smirnov tests of normality results, respectively of 0,171 and 0,129 to regressions 1 and 2, attest for the non normality of residuals.

Based on those results, and inspired by hypotheses H3, a spatial analysis was performed, leading to the development of a spatial auto-regressive model. The results of the spatial analysis are shown in Figure 4. Moran's index, of almost 0.78, suggests a relevant spatial dependence pattern [12] of Household Income in the city of São Paulo. Similar results were obtained for Electric Energy Consumption.

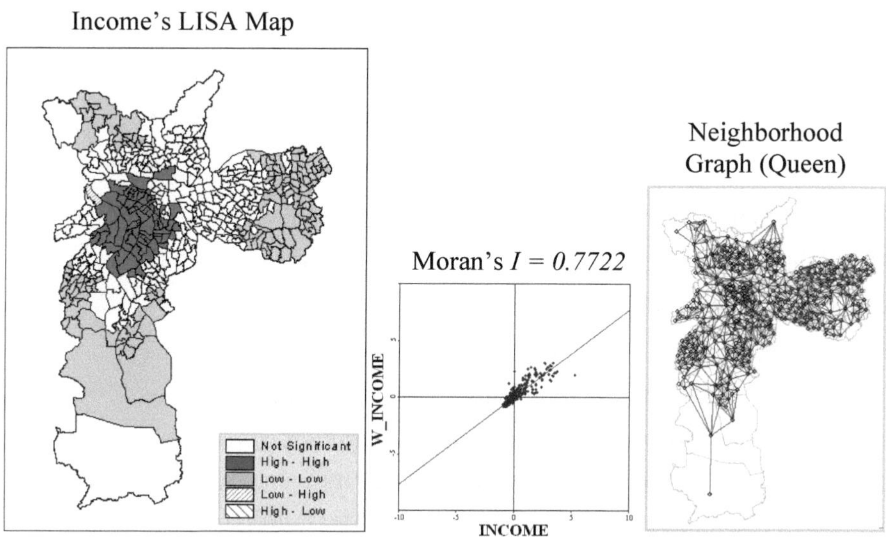

Fig. 3. LISA Map, Moran's scatter plot and indexes of Household Income, using Queen Contiguity Weight

For three different neighborhood matrixes analyzed, the Moran's index showed high values (of about 0.8), in all cases. This suggests a high influence of neighborhood in the Household Income distribution pattern, what is evident in the Moran's scatter plot of Figure 3; there's a trend of income concentration downtown.

A spatial dependence diagnose of household income was then performed by means of a linear regression model. The scales of measure of household income and electricity consumption were changed: their natural logarithms were adopted, instead of their original values. This transformation resulted in a more linear dispersion for income and electricity consumption, and also allowed their relation to be interpreted in terms of elasticity.

An auto-regressive term was also incorporated as a dependent variable in order to normalize the residual errors. Table 5 shows the final autoregressive model, with spatial dependence.

Table 5. Summary of the final Spatial Auto-Regressive Model

Data set : Electricity

Spatial Weight : areaqueen1.GAL (Queen Contiguity Weight)

Dependent Variable : LNINCOME Number of Observations : 456

Mean dependent var : 7.46738 Number of Variables : 3

S.D. dependent var : 0.633242 Degrees of Freedom : 453

Lag coeff. (Rho) : 0.607507

R-squared : 0.936675 Log likelihood : 171.909

Sq. Correlation : - Akaike info criterion : -337.818

Sigma-square : 0.0253932 Schwarz criterion : -325.451

S.E of regression : 0.159352

Estimated Coefficients

Variable	Coefficient	Std.Error	Z-value	Probability
W_LNINCOME	0.6075072	0.02163564	28.07901	0.0000000
CONSTANT	-3.700081	0.1738361	-21.28489	0.0000000
LNCONSUMPTION	1.193655	0.04949749	24.11546	0.000000

The statistics of Log Likelihood and Akaike and Schwarz criteria reveal that the model has an adequate fit. All the estimated coefficients are positive, so hypothesis H1 and H2 are not rejected. Hypothesis H3 can also be examined by means of the spatial model, which increased the R2 coefficient to a level higher than 93%; however, this model was estimated by maximum likelihood, and therefore the R2 coefficient is not calculated the same way as it would be in an ordinary least squares regression, so a comparison in terms of R2 should be avoided.

The residual error of the spatial auto-regressive model presents a normal distribution and a pattern of homoscedasticity. Figure 3 shows this distribution; the Kolmogorov-Smirnov and Breusch-Pagan tests indicate normality and homoscedasticity of the residuals. The conclusion can be graphically verified on the error histogram and on the Moran's scatter plot in Figure 3; the errors do not seem to be auto-correlated at all.

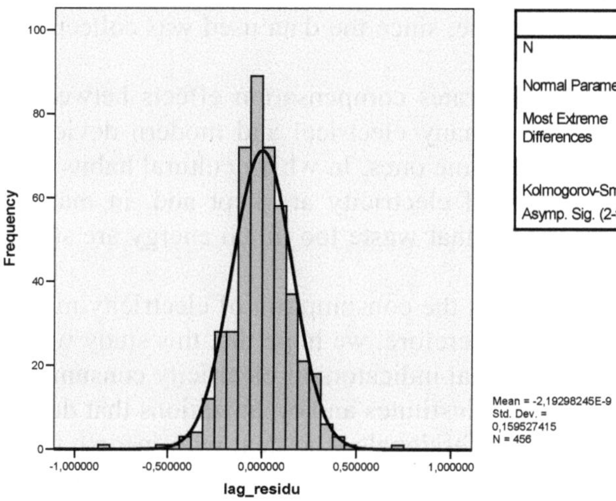

		lag_residu
N		456
Normal Parameters(a,b)	Mean	,0000
	Std. Deviation	,1595
Most Extreme Differences	Absolute	,049
	Positive	,042
	Negative	-,049
Kolmogorov-Smirnov Z		1,052
Asymp. Sig. (2-tailed)		,218

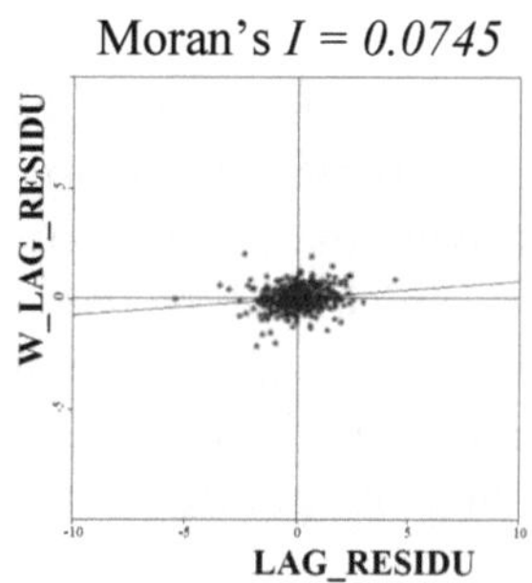

TEST	DF	VALUE	PROB
Breusch-Pagan test	1	0.08303925	0.7732202

Fig. 4. Residual Errors of the Spatial Regression Model

The results of the analysis reveal the existence of a spatially determined pattern of household income distribution in the weighted areas. Furthermore, the consumption of electric energy is shown to be an effective predictor of household income in the studied context.

6 Final Remarks and Managerial Implications

This chapter presented the results of income-predicting models with economic classification and electricity consumption as independent variables. Both economic classification and electricity consumption revealed themselves as proper constructs for predicting income on a territorial basis. The smallest geographic units in most recent researches published in Brazil were districts, so working with the weighted areas can be considered an innovative and more granular approach. The relations found here must be tested with updated data when available, since the data used was collected in 2000.

The use of weighted areas incorporates compensation effects between high income dwellings, which hold many electrical and modern devices that do not waste energy, and low income ones, in which cultural habits of fraud in the levels of consumption of electricity are kept and, in many cases, where old electrical appliances that waste too much energy are still found.

The results in this chapter show that the consumption of electricity may be an efficient predictor of income. Therefore, we hope that this study will motivate the creation of a set of regional indicators of electricity consumption which may be useful for research institutes and organizations that deal with public and urban affairs; and professionals who deal with income estimation in their tasks, such as marketing researchers, policy makers, credit agents and public administration professionals.

References

1. ABEP – Associação Brasileira de Empresas de Pesquisa (2004) Critério de Classificação Econômica Brasil. São Paulo. Available in http://www.anep.org.br. December 22, 2004.
2. Achão C C L (2003) Análise da Estrutura de Consumo de Energia pelo Setor Residencial Brasileiro. Ph.D. Thesis, Universidade Federal do Rio de Janeiro, Rio de Janeiro, Brazil.
3. ANEEL – Agência Nacional de Energia Elétrica (2005) A importância das ações de comunicação para a redução das perdas não-técnicas. Trabalho apresentado no 1° Fórum de Comunicação na Distribuição de Energia Elétrica. Brasília, Brazil.
4. Araújo H P M (1979) O Setor de Energia Elétrica e a Evolução Recente do Capitalismo no Brasil. COPPE-UFRJ. Rio de Janeiro.

5. Bussab W O, Ferreira M (1999) Critério Brasil de Estratificação Socioeconômica: Aspectos Demográficos. In: Proceedings of CLADEA, San Juan, Porto Rico.
6. Cavaretti J L (2005) Consumidor de Baixa Renda e os Desafios das Distribuidoras. In: VIII Encontro Nacional de Conselhos de Consumidores. São Paulo, Brazil.
7. ESRI (2000) ArcView GIS White Paper. Versão 3.2 Available in http://www.esri.com.
8. Francisco E (2002) Customer Franchise – A Mina de Ouro do Geomarketing. InfoGEO, 25, Coluna GEOInsights: 56-57, August. Curitiba: MundoGEO.
9. Francisco E (2004) Análise do Potencial de Uso do Comércio Eletrônico na AES Eletropaulo. Negócios na Era Digital. FGV-EAESP, São Paulo.
10. Francisco E (2006) Relação entre o Consumo de Energia Elétrica, a Renda e a Caracterização Econômica de Famílias de Baixa Renda do Município de São Paulo. Master Thesis, Escola de Administração de Empresas de São Paulo, Fundação Getulio Vargas, São Paulo, Brazil.
11. Governo do Estado de São Paulo (2005) Balanço Energético do Estado de São Paulo 2005. Ano Base 2004. Secretaria de Energia, Recursos Hídricos e Saneamento. São Paulo. Available in http://www.energia.sp.gov.br.
12. Griffith D A (1987) Spatial Autocorrelation – a primer. Washington: Association of American Geographers, 87 pp.
13. Guerreiro A. G, Serra, S T, Carvalho M L R, Silva Filho, M C de (1996) A influência da Venda de Eletrodomésticos na Expansão do Consumo Residencial de Energia Elétrica. Rio de Janeiro: Departamento de Planejamento de Sistemas Energéticos – UNICAMP. In: Anais do VII Congresso Brasileiro de Energia – CBE.
14. IBGE – Instituto Brasileiro de Geografia e Estatística (2000) Censo Demográfico 2000. Características Gerais da População. Rio de Janeiro, Brazil.
15. IBGE – Instituto Brasileiro de Geografia e Estatística (2002) Censo Demográfico 2000. Rio de Janeiro, Brazil.
16. IBGE – Instituto Brasileiro de Geografia e Estatística (2003) PNAD - Pesquisa Nacional por Amostragem de Domicílios: Síntese de Indicadores 2003. Rio de Janeiro, Brazil.
17. IBGE – Instituto Brasileiro de Geografia e Estatística (2004) Pesquisa de Orçamentos Familiares, 2002-2003. Primeiros Resultados. Rio de Janeiro, Brazil: Author.
18. IBGE – Instituto Brasileiro de Geografia e Estatística (2005) Pesquisa Nacional por Amostra de Domicílios, 2004. Rio de Janeiro, Brazil.
19. Madureira R G (1996) Análise dos Aspectos Socioeconômicos, Históricos e Culturais determinantes do Consumo de Energia Elétrica no Setor Residencial Brasileiro. Rio de Janeiro: Departamento de Energia da FEM-UNICAMP.
20. Mattar F N (1996) Porque os Métodos de Classificação Socioeconômicos utilizados no Brasil não funcionam. São Paulo. In: Proceedings of 20° ENANPAD. Available in http://fauze.com.br.

21. Pompermayer M L, Charnet R (1996) Determinantes da Demanda Residencial de Energia Elétrica. In: Proceedings of VII Congresso Brasileiro de Energia – CBE. Rio de Janeiro: Departamento de Planejamento de Sistemas Energéticos – UNICAMP.
22. SEADE – Fundação Sistema Estadual de Análise de Dados (2005) Informações dos Distritos de São Paulo. Access in 12 Apr. 2005 in http://www.seade.gov.br.
23. Silva N L (2004, December 22) Critério Brasil: o mercado falando a mesma língua. São Paulo. Available in http://www.anep.org.br.
24. Wikipedia, The Free Encyclopedia (2006, December 29) Electricity. Access in http://en.wikipedia.org/wiki/Electricity

Shiryaev-Roberts Method to Detect Space-Time Emerging Clusters

Renato Assunção, Thaís Correa

Departamento de Estatística
Universidade Federal de Minas Gerais

1 Introduction

We are interested in monitoring incoming space-time events to detect, as early as possible, an emergent space-time cluster. Assume that point process events (x_i, y_i, t_i) are continuously recorded where $(x_i y_i)$ are the spatial coordinates and t_i is the occurrence time of the *i*-th event. At a certain unknown instant τ, a relatively small cluster of increased intensity starts to emerge. Its location is also unknown. The aim is to let make an alarm go off as soon as possible after τ. The alarm system should also provide an estimate of the cluster location. The alarm system should take into account purely spatial and purely temporal heterogeneity.

In this work we propose a space-time surveillance system with these specifications. It does not require the specification of the spatial pattern or the temporal pattern. It is based on a martingale approach. We detail its theoretical foundation and the corresponding algorithm. Due to lack of space, we study its efficiency elsewhere.

Epidemiological surveillance systems include early statistical warning methods that aim to provide information which can be acted upon to help in the prevention and control of diseases. There is a renewed interest on the development of statistical systems that include spatially referenced information sources due, among other reasons, to heightened concerns about bioterrorism.

The requirements of a surveillance system accounting for spatial structure are generally structured around a basic trade-off: the need for quickly detecting possible outbreaks and epidemics must be balanced against the need for not triggering alarm signals too often unnecessarily.

In this paper, we describe a method to analyze space and time surveillance data in the form of point processes. We propose a probability model to describe eventually emerging spatial clusters with a minimum requirement of user-defined parameters. Based on this model for the emerging spatial clusters, we use the Shiryaev-Roberts statistic and adopt a martingale approach to derive the test properties. Hence, we are able to control the average length run of our surveillance method under the absence of emerging spatial clusters. We define appropriately the average run length for the situation when there are clusters present in the data and illustrate the method in practice. The algorithm is implemented in a freely available stand-alone software and it is expected soon to be in TERRAVIEW.

2 Literature Review

The traditional methods for space-time cluster detection are retrospective in nature. That is, they search in a a database of past events for evidence of clusters' presence. In contrast, our interest is on prospective methods: an events' database is updated regularly and then an algorithm should run to help deciding on the emergence of localized space-time clusters. Hence, the clusters must be alive in the sense that at least some of the most recent events belong to the eventually detected clusters. This brings several difficult problems well known in the artificial intelligence literature: repeated significance tests (at least one every time the database is updated); trade-off between setting up the system to go off as soon as possible after a localized space-time cluster starts to emerge and, at the same time, requiring that the false alarms frequency be kept at a minimum.

A thorough literature review can be found in the book [4] or in [6]. We give here a very brief overview of the main proposals. There are non-spatial methods derived from quality control ideas concerned with monitoring a stochastic process on time. The Shewart Chart Control is a very simple and popular method but it is not sensitive to small changes in the process. The Cummulative Sum (CUSUM) method accumulates the recent evidence to the previous data to trigger a threshold limit. It has been shown that it has optimal properties in very simple scenarios. Exponentially weighted moving average also accumulates evidence, as the CUSUM method, but it discounts observations as they get old. All these methods assume data are independent in time, not a realistic assumption. [1] uses a Shiryaev-Roberts statistics to allow for dependent data.

There are few space-time oriented proposals. Two recent and promising ones are [2], who proposed a space-time scan statistic for areal data, and [3] and [5], that suggested a statistic based on local Knox statistic.

We introduced a new method focusing on point process data. That is, there is no risk population info. The null hypothesis of interest is that we have a separable events density with unspecified and arbitrary spatial and temporal heterogeneity. As alternative, we assume that somewhere, at some moment, few localized space-time high intensity clusters start to emerge. We develop a likelihood model for this pair of hypotheses and monitor the incoming data with a spatial version of the Shyriaev-Roberts statistic.

3. Basic Concepts and notation

The Shiryayev-Roberts method was developed for temporal processes only. Suppose that a sequence of possibly dependent random variables $X_1, X_2, \ldots$ is observed. Let $f_{(k)}(x_1, x_2, \ldots, x_n)$ be the joint density distribution of the first n random variables when a cluster starts to emerge at moment $\tau = k$. When no cluster ever emerges, we write $f_{(\infty)}(x_1, x_2, \ldots, x_n)$. Any surveillance method implies a stopping time N, the first moment when the alarm goes off.

Let $E_{(k)}(\cdot)$ be the expectation with respect to $f_{(k)}(x_1, x_2, \ldots, x_n)$. $E_{(\infty)}(N)$ is called the Average Run Length and it is denoted by ARL^0. Clearly, it is desirable to keep ARL^0 small and, for that, the user establishes an acceptable minimum threshold B for this parameter. That is, we want $ARL^0 = E_\infty(N) > B$.

The Shiryayev-Roberts test statistic is given by

$$R_n = \sum_{k=1}^{n} \frac{f_{(k)}(X_1, X_2, \ldots, X_n)}{f_{(\infty)}(X_1, X_2, \ldots, X_n)}$$

The alarm goes off if R_n is too large, that is, if $R_n > A$. The stopping time is N_A: the alarm goes off by the first time at N_A where

$$N_A = \min\{\, n \mid R_n \geq A \,\}$$

It remains to find A such that $ARL^0 = E_\infty(N) > B$.

Under P_∞, the sequence

$$\Lambda_{k,n} = \frac{f_{(k)}(X_1, X_2, \ldots, X_n)}{f_{(\infty)}(X_1, X_2, \ldots, X_n)}$$

is a martingale with expected value equal to 1 (even with dependent observations). Therefore,

$$R_n - n = \sum_{k=1}^{n} (\Lambda_{k,n} - 1)$$

is a zero mean martingale. By the Optional Sampling Theorem, we have

$$E_\infty(R_{N_A} - N_A) = 0 \Rightarrow E_\infty(R_{N_A}) = E_\infty(N_A)$$

By definition, $R_{N_A} \geq A$ and hence $E_\infty(N_A) \geq A$. Therefore, taking $A=B$ satisfies the condition $ARL^0 = E_\infty(N) > B$.

There are several advantages associated with the Shiryayev-Roberts (SR) method. First, it can be shown that it exhibits some optimal properties in some simple scenarios. Furthermore, in terms of the delay time for the alarm going off after the purely temporal clusters strats to emerge, the SR and CUSUM are similar. The SR method does not require independence between observations. And it can also be shown that SR is at least as efficient as some optimal classical procedures.

The major disadvantage of the SR method is that it depends on the complete specification of the joint distribution of the sequence $X_1, X_2, \ldots, X_n$ after a change occurs at $\tau = k$. If this is difficult to be done in the purely temporal context, in the space-time situation it seems hopeless. However, we found a way out as we explain next.

3.1 Our proposal for space-time clusters

Let N be a Poisson process in $\Re^3$ partially observed in the three-dimensional region $A\times[0,T]$. Let $N(C_i)$ be the number of events in the cylinder C_i. The random variable $N(C_i)$ is distributed according to a Poisson distribution with mean $\mu(C_i)$ and $\mu(C_i)$ is unknown.

Let $\lambda(x,y,t)$ be the intensity function of the events in $A\times[0,T]$. Consider a cylinder C_i in $\Re^3$ (see Figure 1) and let $\mu(C_i)$ be the integral over C_i of the intensity $\lambda(x,y,t)$, while μ is the expected number of events in all the region $A\times[0,T]$. Define the marginal spatial and tempo

$$\lambda_S(x,y)=\mu^{-1}\int_{[0,T]}\lambda(x,y,t)\,dt$$

and

$$\lambda_T(t)=\mu^{-1}\iint_A\lambda(x,y,t)\,dx\,dy\,.$$

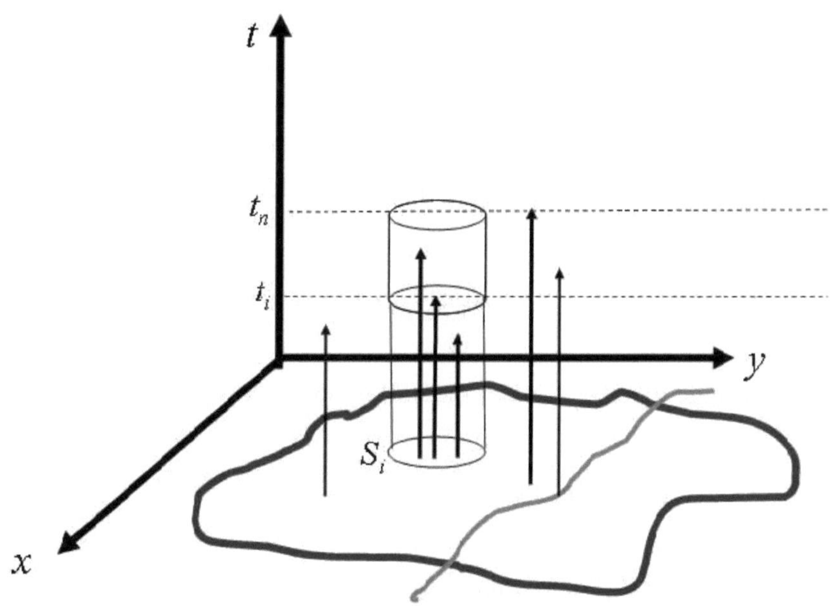

Fig. 1. The typical 3-dim cylinder and some events

We define now the pair of hypotheses. The null hypothesis (no cluster scenario) is established as a separable intensity

$$\lambda(x,y,t) = \mu \lambda_S(x,y) \lambda_T(t)$$

where $\lambda_S(x,y)$ and $\lambda_T(t)$ are arbitrary and unspecified. That is, they are nuisance parameters. The alternative hypothesis assumes that there exists a time τ, a constant ε>0, and a cylinder C_τ (yet to be defined) such that

$$\lambda(x,y,t) = \mu\, \lambda_S(x,y) \lambda_T(t) \left(1 + \varepsilon I_C(x,y,t)\right)$$

The parameter ε is the relative change on the events intensity within the cluster and it must be user-specified.

To define a useful class of cylinders C_τ, we start considering that, if a higher incidence cluster emerges, we must be able to detect it through the observed events. That is, non-events (or void spaces) do not bring information about an emerging cluster. Hence, we decided to constrain τ to be equal to one of the observed t_i's; the cylinders should be in the form of a circle *S* times a temporal interval. The time interval is $[t_i, t_n]$ where t_n is the last event, since interest is only in alive clusters. The cylinder *S* has a radius ρ specified by the user.

We can now proceed to determine the mean $\mu(C_i)$. From the non-homogeneous Poisson process properties, under the null hypothesis, we have:

$$\mu(C_i) = \int_{C_i} \lambda(x,y,t)\, dx\, dy\, dt = \mu \int_{S_i} \lambda_S(x,y)\, dx\, dy \int_{[t_i,t_n]} \lambda_T(t)\, dt$$

An estimate of $\mu(C_i)$ under the null hypothesis is given by

$$\hat{\mu}(C_i) = \frac{N(S_i \times [0,T])\, N(A \times [t_i, t_n])}{n}$$

where $N(S_i \times [0,T])$ is the number of events within circle S_i irrespective of time; $N(A \times [t_i, t_n])$ is the number of events with times between t_i and t_n, irrespective of spatial location; and *n* is the total number of events (see Figure 1).

To define the test statistic, we consider the likelihood of space-time Poisson processes. Under the null hypothesis, we have

$$L_\infty = \left(\prod_{i=1}^{n} \lambda(x_i, y_i, t_i)\right) \exp\left(-\int_{A\times[0,T]} \lambda(x,y,t)\,dx\,dy\,dt\right)$$

Under the alternative, we have

$$L_\tau = \left(\prod_{i=1}^{n} \lambda(x_i, y_i, t_i)\left(1 + \varepsilon I_{C_\tau}(x_i, y_i, t_i)\right)\right) \exp\left(-\int_{A\times[0,T]} \lambda(x,y,t)\,dx\,dy\,dt\right)$$
$$\exp\left(-\varepsilon \int_{C_\tau]} \lambda(x,y,t)\,dx\,dy\,dt\right)$$

where $\lambda(x,y,t) = \mu\, \lambda_S(x,y)\, \lambda_T(t)$ and C_τ is the putative cluster cylinder.

Therefore, a space-time version of the SR test statistic R_n becomes

$$\begin{aligned} R_n &= \sum_{\tau=1}^{n} \frac{L_\tau}{L_\infty} \\ &= \sum_{\tau=1}^{n} \left\{ \left[\prod_{i=1}^{n} \left(1 + \varepsilon I_{C_\tau}(x_i, y_i, t_i)\right)\right] \exp\left(-\varepsilon \int_{C_\tau} \lambda(x,y,t)\,dx\,dy\,dt\right)\right\} \\ &= \sum_{\tau=1}^{n} (1+\varepsilon)^{N(C_\tau)} \exp(-\mu(C_\tau)) \end{aligned}$$

with $\mu(C_\tau)$ estimated as explained before.

The parameter $\varepsilon > 0$ is known (user-specified) and measures the anticipated relative change in the events' density. Our surveillance method calculates R_{n+1} as the $n+1$-th event arrives with $\hat{\mu}(C_\tau)$ rather than $\mu(C_\tau)$. The alarm goes off when $R_n > A$ for the first time.

In summary, the algorithm associated with our proposal needs as input: n cases events given by the coordinates x, y, and time t; the value of three tuning parameters: ε, the anticipated relative change in density within the cluster; the anticipated radius ρ for the cluster; the threshold A, which should be approximately equal to the desired ARL^0. Iteratively in n, calculate R_n. The output is a sequence of values R_n where n is the number of events. If $R_n > A$ for any n, the alarm goes off.

4 Illustration

As an illustration, we used a classical example of retrospective detection of space-time clustering, the locations of cases of Burkitt's lymphoma in the Western Nile district of Uganda in the period from 1960 to 1975, originally studied in [7]. The time variable is recorded as the number of days starting from an origin of 1 Jan 1960. [5] found evidence of space-time clusters using local Knox tests and adopting a probability of false alarm of 0.1. However, we have not been able to reproduce his results using his methods.

Fig. 2. Burkitt lymphoma cases in Uganda.

The tuning parameters in our surveillance method were:

- We fix $\varepsilon = 0.5$, a large anticipated change.
- $\rho = 210\ km$ (weighted average of values used by Rogerson (2001)).
- Limit A of the alarm equal to 161. In average, we expect 161 events before the alarm goes off without need.

Figure 3 shows the R_n versus n. We can observe that the alarm goes off at event number 148 (February, 1973). Typically, there was little variation of the detected space-time cluster over many different tuning parameter choices. One pattern we found is that, for ρ= 2.5, 5, 10, 20 km, the smaller ε, the longer it takes for the alarm to go off.

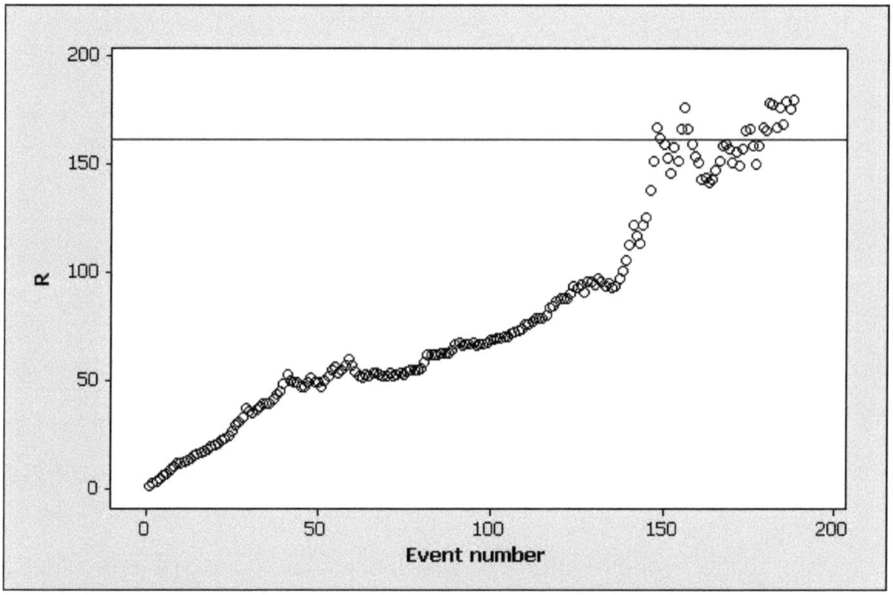

Fig. 3. Graph of R_n versus n for the Burkitt lymphoma dataset.

References

[1] Kennett, R., and Pollak, M. (1996) Data-analytic aspects of the Shiryaev-Roberts control chart: surveillance of a non-homogeneous Poisson process. Journal of Applied Statistics, 23, 125–137.

[2] Kulldorff, M. (2001) Prospective time periodic geographical disease surveillance using a scan statistic. Journal of the Royal Statistical Society, Series A, 164:61–72.

[3] Kulldorf, M., Heffernan, R., Hartman, J., Assunção, R., and Mostashari F. (2005) A space-time permutation scan statistic for disease outbreak detection. PLoS Medicine, 2, 216–224.
[4] Lawson, A. B., and Kleinman, K. (2005) Spatial and Syndromic Surveillance for Public Health. John Wiley & Sons, New York.
[5] Rogerson, P. A. (2001) Monitoring point patterns for the development of space-time clusters. Journal of the Royal Statistical Society, Series A, 164: 87–96.
[6] Sonesson C., and Bock D. (2003) A review and discussion of prospective statistical surveillance in public health. Journal of the Royal Statistical Society. Series A, 166, 5–21.
[7] Williams, E. H., Smith, P.G., Day, N. E., Geser, A., Ellice, J., and Tukei, P. (1978) Space-time clustering of Burkitt's lymphoma in the West Nile district of Uganda: 1961-1975. British Journal of Cancer, 37(1): 109-122.

Testing association between origin-destination spatial locations

Renato Martins Assunção, Danilo Lourenço Lopes

Departamento de Estatística
Universidade Federal de Minas Gerais

1 Introduction

Point processes are probabilistic frameworks to analyze spatial patterns composed by random point features, called events, stored in GIS database. In a multivariate point process, the events are of two or more different types such as the locations of disease cases and a set of locations labelled as control individuals or as the positions of plants in a planar region labelled according to their species [2]. Usually, the spatial analysis of multivariate point processes is concerned with two questions. The first one concentrates on the comparison between the individual patterns of the component processes. Typically, the interest is to decide if one spatial pattern (such as the disease cases) has some degree of spatial clustering with respect to another spatial pattern (such as the controls' pattern) (see [5]), perhaps identifying some putative sources of increased relative intensity ([3]; [4]). The second question concentrates on testing the independence of two (or more) point patterns and therefore attention is directed to the *joint* distribution of the processes [7]. It is common, for example, to test if the presence of an event of a certain type in a location either inhibits or stimulates the nearby presence of events of other types.

In this paper, we are interested in another type of situation that is analyzed less often in spatial studies: bivariate point processes structured as origin-destination locations. To illustrate, consider the locations of m = 6339 car thefts occurred in he years 2000 and 2001 in Belo Horizonte, a 2 million inhabitants Brazilian city (see left hand side map in Figure 1). Some of the stolen cars are never retrieved by the police or they are retrieved outside the city boundary. The right hand side map of Figure 1 shows the locations of the $n = 5257$ cars eventually found within the city limits. Conditioning on the car retrieval, the interest is to know if there is some type of spatial dependence between the two locations of each stolen car.

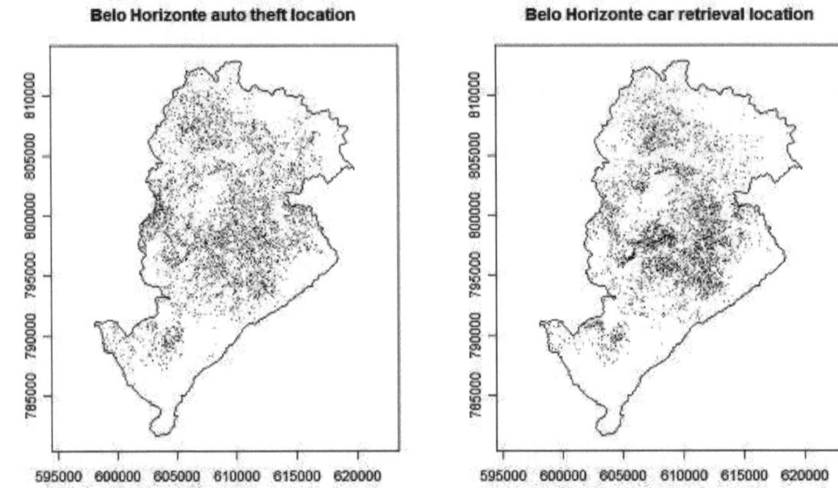

Fig. 1. Illustrating maps of auto theft locations and car retrieval in Belo Horizonte from Aug 2000 to Aug 2001

Other possible applied settings that generate bivariate linked point processes are: migration between different regions, murder location and the victim's residence address, origin-destination survey data from transportation studies.

These situations have in common that two (or more) spatial point patterns have linked events. That is, for each event in one point pattern, there is one (or more) corresponding events in the other point pattern. In this paper, we concentrate on bivariate patterns, one of them called origin process, and the other called destination process. We denote this kind of data as origin-destination point patterns or bivariate linked point processes.

We introduce a new correlation measure for origin-destination point patterns and we point its similarities with Knox's statistic, a common tool in spatial-temporal studies. Our measure of correlation tests if pairs of events that are close in origin tend to be close in destination too. Our proposal has some attractive features: it is simple to understand, it is easy to be calculated, and it has an asymptotic distribution that does not depend on the spatial-temporal pattern of the data. In particular, it can be applied to any type of spatial point pattern. Our measure derives its theoretical properties from the fact that it is based on a score score test statistic, the locally most powerful test within a certain probabilistic model.

We proceed as follows. Section 2 proposes a model to bivariate linked spatial point data. Section 3 develops the locally most powerful test for independence between origin-destination hypothesis. Section 4 describes an application to car theft-retrieval dataset in Belo Horizonte and Section 5 presents our conclusions.

2 A stochastic model for bivariate linked point processes

The main origin-destination point pattern motivation for this paper is composed by the locations of car thefts and the locations of their eventual retrieval in Belo Horizonte witin a year. In Figure 2, the map in the right hand side shows all the linked locations of the 5257 vehicles stolen between August 2000 and August 2001. In the left hand side, only a random sample of 50 stolen vehicles in Belo Horizonte is shown with each arrow going from the theft position to its corresponding retrieval position.

Suppose that a car is stolen at position $\boldsymbol{x}$, corresponding to the origin of the event. Consider the probability distribution of a stolen car retrieval location $\boldsymbol{y}$, given that it has been stolen at location $\boldsymbol{x}$. Let $f(\boldsymbol{y}|\boldsymbol{x})$ be the density of the destination $\boldsymbol{y}$ given that the origin is $\boldsymbol{x}$. For each possible origin $\boldsymbol{x}$ we have a surface $f(\boldsymbol{y}|\boldsymbol{x})$ showing the most likely destinations of events originating at $\boldsymbol{x}$.

Before embarking on estimation of a stochastic model for $f(\boldsymbol{y}|\boldsymbol{x})$, it is worth to verify if the hypothesis of independence between origin-destination point patterns holds. This is a minimum requirement for such modeling. If there is no evidence for dependence, there is no point on estimating $f(\boldsymbol{y}|\boldsymbol{x})$. Hence, a first step in the analysis is to test if $f(\boldsymbol{y}|\boldsymbol{x})$ is the same, whatever the location $\boldsymbol{x}$. If this is so, there is origin and destination locations are independent. Otherwise, we say that there is dependence or association between origin and destination locations.

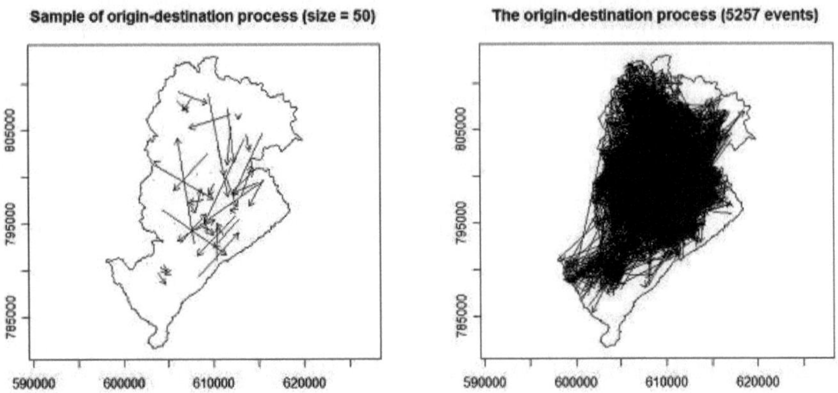

Fig. 2. Sample (left) and entire dataset (right) of thefts and retrievals in Belo Horizonte. Each arrow goes from a theft location to the correspondent retrieval location.

Let (N_1, N_2) be a bivariate linked point process observed in a finite polygon $A \subset R^2$. The available data are a set of n events (x_i, y_i), $i=1,...,n$, composed by pairs of spatial locations in A. The first location $\boldsymbol{x}_i = (x_{1i}, x_{2i})$ is called the origin event and the second location $\boldsymbol{y}_i = (y_{1i}, y_{2i})$ is the destination event. We denote by φ the unordered set $\{(x_1, y_1), ..., (x_n, y_n)\}$ of pairs of locations.

Conditioning on the total number n of observed events, we can work with the probability density distribution $p(\varphi)$ of the n events in A^{2n}, which must be invariant with respect to the ordering of the locations ([2], page 123-125). Under the assumption of independence between origin and destination locations, this probability density can be written as

$$p(\varphi) = C \exp(g(x_1, \cdots, x_n) + h(y_1, \cdots, y_n)) \tag{1}$$

where C is a normalizing constant. The function g models the interaction among the origin events as well as any spatial variation in the first order intensity of this marginal process. The function h has the same role with respect to the destination events. The functions g and h can be chosen arbitrarily as long as the normalizing constant in Eq. (1) is finite and as long as each one of them is invariant with respect to all permutations of the events labels.

To introduce some kind of interaction between origin and destination, we propose a model with pairwise interaction functions. Namely, we assume that

$$p(\varphi) = C \exp\left(g(x_1, \cdots, x_n) + h(y_1, \cdots, y_n) - \sum_{i<j} \phi((x_i, y_i), (x_j, y_j); \theta) \right) \tag{2}$$

where ϕ is a potential function depending on a set of parameters $\boldsymbol{\theta}$. One simple special case is to define two positive thresholds r_x and r_y and then Eq. (2) becomes

$$\begin{aligned} p(\varphi) &= C \exp\left(g(x_1, \cdots, x_n) + h(y_1, \cdots, y_n) - \theta \sum_{i<j} I\left[\left| x_i - x_j \right| < r_x\right] I\left[\left| y_i - y_j \right| < r_y\right] \right) \\ &= C \exp\left(g(x_1, \cdots, x_n) + h(y_1, \cdots, y_n)) - \theta T(\varphi) \right) \end{aligned} \tag{3}$$

where $I[.]$ is the indicator function and $T(\varphi)$ is the number of pair of events that are within the threshold limits at the origin and at the destination.

The case $\theta = 0$ corresponds to the independence between origin and destination processes. If $\theta < 0$, the result is an inhibition process: pairs of nearby events at the origin will tend to be farther apart in the destination. Values of $\theta > 0$ correspond to the spatial clustering of the pair of origin-destination events.

In principle, the functions g and h in Eq. (2) can be quite general. Isolated analysis of each one of the patterns, origin events and destination events, can provide parametric or non-parametric estimates of these functions under the null hypothesis. However, when the main interest is to test for the presence of spatial correlation between the two types of events, it is useful to condition on the observed marginal locations. This eliminates the nuisance functions g and h.

Let $\pi_1, \cdots, \pi_n$ be a permutation of the indexes $1, \cdots, n$. The probability density distribution for the unlinked and unordered locations is denoted by $p(\{\mathbf{x}_1, \ldots, \mathbf{x}_n\}, \{\mathbf{y}_1, \ldots, \mathbf{y}_n\})$ and given by

$$C \exp(g(x_1,\cdots,x_n) + h(y_1,\cdots,y_n)) \quad (4)$$
$$* \sum_{\pi} \exp\left(-\sum_{i<j} \phi((x_i, y_{\pi_i}),(x_j, y_{\pi_j});\theta) \right)$$

where the sum is taken over all *n!* possible permutations.

Therefore, the distribution of φ conditional on the unordered set of origin locations $\{x_1, ..., x_n\}$ and on the unordered set of destinations $\{y_1, ..., y_n\}$ is given by

$$p(\varphi \mid \{x_1,\cdots,x_n\},\{y_1,\cdots,y_n\}) = \quad (5)$$
$$\frac{\exp\left(-\sum_{i<j} \phi((x_i, y_i),(x_j, y_j);\theta)\right)}{\sum_{\pi} \exp\left(-\sum_{i<j} \phi((x_i, y_{\pi_i}),(x_j, y_{\pi_j});\theta)\right)}$$

The hard to calculate normalizing constant C is substituted by a simpler normalizing constant although its sum over *n!* terms is computationally demanding unless n is unrealistically small.

To simplify notation, we let $\pi\varphi$ to denote the unordered set of events with the destinations shuffled by some arbitrary permutation $\pi_1,\cdots,\pi_n$:

$$\pi\varphi = \{(x_1, y_{\pi_1}),\cdots,(x_n, y_{\pi_n})\}$$

3 Testing for spatial correlation

In the remaining of the paper, we focus our attention in the model of Eq. (3). For the conditional distribution in Eq. (5), the log-likelihood of the interaction parameter θ is given by

$$l(\theta) = -\theta T(\varphi) - \log\left(\sum_{\pi} \exp(\theta T(\pi\varphi)) \right)$$

It is clear that $T(\varphi)$ is a natural sufficient statistic for the parameter θ. The score statistic is then given by

$$\frac{\partial l}{\partial \theta} = -T(\varphi) + \frac{\sum_{\pi} T(\pi\varphi)\exp(\theta T(\pi\varphi))}{\sum_{\pi} \exp(\theta T(\pi\varphi))} \tag{6}$$
$$= E_{\pi,\theta}[T(\pi\varphi)] - T(\varphi),$$

a contrast between observed and expected values of the sufficient statistic, where the expectation of $T(\pi\varphi)$ is taken with probabilities given by all *n!* normalized values

$$\frac{\exp(\theta T(\pi\varphi))}{\sum_{\pi} \exp(\theta T(\pi\varphi))}.$$

The score test statistic is given by

$$\left.\frac{\partial l}{\partial \theta}\right|_{\theta=0} = \frac{1}{n!}\sum_{\pi} T(\pi\varphi) - T(\varphi). \tag{7}$$

The moments of $T(\varphi)$ under the null hypothesis are easily obtained. With the origin and the destinations fixed and taking the expectation under all permutations of the destination indexes, we have the following for the expected value:

$$E_{\pi}[T(\pi\varphi)] = E_{\pi}\left(\sum_{i<j} I[|x_i - x_j| < r_x] I[|y_{\pi_i} - y_{\pi_j}| < r_y]\right)$$
$$= \sum_{i<j} I[|x_i - x_j| < r_x] E_{\pi}(I[|y_{\pi_i} - y_{\pi_j}| < r_y])$$

because of the independence between origin and destination and the conditioning on the marginal patterns. There are $n(n-1)/2$ pairs of events and n_y of them are close to each other at the destination. Hence, the probability that the pair π_i, π_j of a random permutation is one of them is $2n_y / n(n-1)$ and therefore

$$E_{\pi}[T(\pi\varphi)] = \frac{n_y}{\binom{n}{2}} \sum_{i<j} I[|x_i - x_j| < r_x] = \frac{n_y n_x}{\binom{n}{2}}$$

where n_x is the number of pairs of origin events that are close to each other.

The second moment is given by

$$E_\pi\left\{[T(\pi\varphi)]^2\right\} = E_\pi\left(\sum_{i<j}\sum_{k<l} I\left[\left|x_i - x_j\right| < r_x\right] I\left[\left|x_k - x_l\right| < r_x\right] I\left[\left|y_{\pi_i} - y_{\pi_j}\right| < r_y\right] I\left[\left|y_{\pi_k} - y_{\pi_l}\right| < r_y\right]\right)$$

$$= \sum_{i<j}\sum_{k<l} I\left[\left|x_i - x_j\right| < r_x\right] I\left[\left|x_k - x_l\right| < r_x\right] E_\pi\left(I\left[\left|y_{\pi_i} - y_{\pi_j}\right| < r_y\right] I\left[\left|y_{\pi_k} - y_{\pi_l}\right| < r_y\right]\right)$$

also conditioning on the marginal patterns. Following, we present the development of such expectation for each of the cases can arise. First, for $i = k, j = l$ we have:

$$\sum_{i<j}\left(I\left[\left|x_i - x_j\right| < r_x\right]\right)^2 E_\pi\left\{\left(I\left[\left|y_{\pi_i} - y_{\pi_j}\right| < r_y\right]\right)^2\right\} = \frac{n_y n_x}{\binom{n}{2}}$$

as demonstrated previously. For the case where only one of *(i,j)* is equal to one of *(k,l)*: there are $\frac{n}{2}\binom{n-1}{2}$ pairs of pair cases with one case in common. n_{2y} of them are composed by pairs of 'case pairs that are close to each other in destination' that have one case in common. Hence the probability that the quadruplet $\pi_i, \pi_j, \pi_l, \pi_k$ is one of them is $4n_{2y} / n(n-1)(n-2)$ and therefore:

$$\frac{4n_{2y}}{n(n-1)(n-2)}\sum_{i<j}\sum_{k<l} I\left[\left|x_i - x_j\right| < r_x\right] I\left[\left|x_k - x_l\right| < r_x\right] = \frac{4n_{2y}n_{2x}}{n(n-1)(n-2)}$$

where n_{2y} is the number of pairs of origin events pairs that are close to each other and have one event in common.

Finally, for the case where *i,j,k,l* are all different among them, there are $\binom{n}{2}\binom{n-2}{2}$ possible quadruplets. The number of them that are close to destination in both pairs is the number of crossing products of the n_y pairs, minus the number of repeated pairs, minus those ones that have one case in common, which gives to us the quantity of n_y^2 - n_y - n_{2y}. So, the expectation for this case is given by:

$$= \frac{4\left[n_y(n_y-1)-n_{2y}\right]}{n(n-1)(n-2)(n-3)} \sum_{i<j} \sum_{\substack{k<l \\ k \neq i,j \\ l \neq i,j}} I\left[\left|x_i - x_j\right| < r_x\right] I\left[\left|x_k - x_l\right| < r_x\right]$$

$$= \frac{4\left[n_y(n_y-1)-n_{2y}\right]\left[n_x(n_x-1)-n_{2x}\right]}{n(n-1)(n-2)(n-3)}$$

Uniting all possible cases, we have:

$$E_\pi\left\{\left[T(\pi\varphi)\right]^2\right\} = \frac{n_y n_x}{\binom{n}{2}} + \frac{4n_{2y}n_{2x}}{n(n-1)(n-2)} + \frac{4\left[n_y(n_y-1)-n_{2y}\right]\left[n_x(n_x-1)-n_{2x}\right]}{n(n-1)(n-2)(n-3)}$$

Hence, an asymptotic test based on the normal approximation to the score test statistic can be carried out. Alternatively, we can use a Monte Carlo test, which do not require asymptotic arguments and which provides exact p-values [1]. By sampling a large number *B-1* of permutations independently and with equal probability, the p-value is given by *(1+k)/B* where *k* is the number of sampled values $\left|T(\pi\varphi)\right|$ greater or equal than the observed value of $\left|T(\varphi)\right|$.

$T(\pi\varphi)$, the score test statistic for origin-destination independence test, is similar to another well-known test statistic, the Knox test statistic (see [6]), proposed for testing space-time interactions. Specifying a spatial and a temporal critical distance, it is possible to indicate when a pair of events is close in space or close in space and in time. Knox test statistic X is defined as the number of pairs of events that are simultaneously close in space and in time. A large number X would be an indication that cases that are close in space tend also to be close in time leading to space-time interaction. In fact, the score test statistic $T(\varphi)$, that was derived from a simplifying proposal on the Gibbs process with parwise interaction functions, may be also seen as an extension of Knox test statistic, used for spatial-temporal data, to origin-destination data.

4 Application

For illustration, we use the auto theft and car retrieval locations data from a large Brazilian city, Belo Horizonte, during the period from August 2000 to August 2001 collected by the Military Police of Minas Gerais based on their police records of crime events. From the 6339 vehicles listed in the database, 5257 of them are georeferenced by both of its theft and retrieval locations, 653 of them are georeferenced in destination but not in origin, 268 of them were stolen and not retrieved and 161 of them are not georeferenced on neither one of theft and retrieval locations.

Table 1 presents the results of applying the score test in our dataset with six distance thresholds, the same for both, origin and destination. We used 3999 Monte Carlo simulations to carry out the test. For each distance threshold, we present the values of the observed statistic, as calculated with the real dataset, the expectation of the test statistic under independence and the Monte-Carlo p-value. As seen in Table 1, the hypothesis of independence between origin and destination is rejected at a significance level of 5% in all threshold levels.

$r_x = r_y$	$T(\varphi)$	$E[T(\varphi)]$	P-value
750 m	13,884	1,920	2.5E-4
1,500 m	93,262	21,966.32	2.5E-4
2,500 m	371,307	129,216.81	2.5E-4
3,250 m	734,011	314,244.47	2.5E-4
4,000 m	1,241,616	622,437.05	2.5E-4
5,000 m	2,115,013	1,240,667.53	2.5E-4

Table 1. Score test results using different thresholds.

As there is evidence of dependence between origins and destinations, one further step in the analysis is to fit a stochastic model for the bivariate point patterns. An exploratory analysis of our dataset revealed that in Belo Horizonte, if a car is stolen somewhere, the two most likely regions for its retrieval location are the neighborhood of the theft location and the South-Center region of the city. Hence, a simple and useful model for this city is:

- Given that a car is stolen at the origin $\boldsymbol{x}$, its destination tends to be a mixture of two densities.
- With probability $p(\boldsymbol{x})$, it stays around $\boldsymbol{x}$.
- With probability $1\text{-}p(\boldsymbol{x})$, it tends to be attracted towards the South-Center region.

5 Conclusions

In this paper, we introduce a new type of method to analyse bivariate linked point patterns. The interest in om two geographical patterns, the first containing the origin locations, while the second contains the destination pattern of point events in a map. Before any modelling of the association between the patterns, one must first to test if there is any evidence for this association. The main objective of this paper is the introduction of such a test. We showed that a simple statistic, the number of pair of events that are close in both, origin and destination maps, has some optimality properties to test the independence between origin and destination. It is also very simple to calculate and can be applied in virtually any specific bivariate point patterns.

One additional advantage of our proposed test statistic is that it does not require risk population informatioion. Hence, we do not need to know where all the cars are located or are more liked to be stolen to test the origin-destination hypothesis. It has an asymptotic normal distribution with known moments that can be used when simulation is too costly.

Acknowledgements

This research was supported by Conselho Nacional de Desenvolvimento Científico e Tecnológico, CNPq, with grant number 301173/97-4. The authors thank the Military Police of Minas Gerais for providing the analyzed dataset in the paper. We would thank to anonymous referees for their helpful comments.

References

[1] Besag, J. and Clifford, P. (1989) Generalized Monte Carlo significance tests. *Biometrika*, **76**, 633-642.

[2] Daley, D.J., and Vere-Jones, D. (2003) An introduction to the theory of point processes, 2nd edition. New York: Springer-Verlag.

[3] Diggle, P.J. (1990). A point process modelling approach to raised incidence of a rare phenomenon in the vicinity of a pre-specified point. *Journal of the Royal Statistical Society A*, **153**, 349-362.

[4] Diggle, P.J. and Rowlingson, B. S. (1994). A conditional approach to point process modelling of raised incidence. *Journal of the Royal Statistical Society A*, **157**, 433-440.

[5] Kelsall, J. and Diggle, P.J. (1995). Kernel estimation of relative risk. *Bernoulli*, **1**, 3-16.

[6] Knox, G. (1964). Epidemiology of childhood leukemia in Northumberland and Durham. *Brit. J. Prevent. & Social Med.*, **18**, 17-24.

[7] Lotwick, H. W. and Silverman, B. W. (1982) Methods for analyzing spatial processes of several types of points. *Journal of the Royal Statistical Society B*, **44**, 406-413.

GIS Development for Energy Distribution Network Restoration with an Integrated Interface

Nelkis de la Orden Medina, Rodrigo Lapa, Marcelo Antonio Nero, Ricardo Luis Guimarães dos Santos, Antonio Valério Netto.

Cientistas Associados Ltda.

1 Introduction

The power distribution companies are being requested on the quality of the power supply and its distribution [2] [3]. However, power outages are to be expected when there is a fault in the energy distribution network. In this case, it is necessary to de-energize the power lines where there is a fault and reconfigure the system in order to minimize the impact over the consumers in a process known as power restoration [6].

The power distribution systems have strategically located switches along the circuits which allow the system reconfiguration through the restoration of load among nearby circuits, in the same substation or in circuits of other substations, in case of a fault in the system.

Although this functionality is present in several SCADA systems (*Supervisory Control and Data Acquisition*), it is common the use of drawings that do not truly represent the energy distribution network and its location, preventing fast and precise measures. The tool that allows this functionality is the GIS (*Geographic Information System*).

According to Stoter and Zlatanova [9] GIS has become a sophisticated system for storing and analyzing spatial and thematic database since the early 90's. However, 2D GIS is rather limited concerning some types of analysis (noise propagation models, geological models, air pollution models, among others), which can be improved with the use of 3D GIS. Nevertheless, the 3D GIS development is much more complex [12] due to the requirement of the cartography and the detailed information, increasing the costs sensibly. According to Zhu et al [11] one of the main problems of this system is the data and format heterogeneity. Rahman [8] claims that there have been great improvements on the presentation and the analysis surface of commercial software programs and that the greatest challenge today is the topological structure consolidation of the 3D GIS [13].

With the use of software programs that simulate 3D spaces, it is possible to have a faster and more precise interpretation of a greater amount of information. This enables a better understanding of the existing spatial relations among the analyzed elements and also facilitates the visualization of complex situations, which could only be possible through a great amount of maps and documents.

In this context, it is suggested the use of an open source 3D GIS application and the development of a new software program in order to reconfigure and restore the energy distribution network.

2 Conception

Nowadays several energy distribution companies make use of the unifilar diagram of the energy distribution network over a printed map in the restoration process. From this analysis, a new configuration is proposed and the switches to be changed are defined. After that, computer programs that simulate the new power distribution configuration are applied and in case the reconfiguration is not appropriate a new study is made with the energy distribution system [4].

Apart from the technical criteria for the power restoration, other factors should be considered such as the geographical location of the fault and the affected consumers. In this sense, it is essential to minimize the amount of consumers affected in order of priority in which the power outage must be avoided (hospitals, plants, fire departments, etc). Also, data such as the location of railway stations, rivers and bridges are important factors in the operation of a power distribution system [4].

Aiming at offering a better visualization of the power distribution grid in a contextualized way (such as the navigation system, location of electrical elements, etc) it was proposed the creation of the ENS3D (*Energy Network System 3D*) which integrates an intelligent system with evolutive algorithm with a virtual environment in a GIS platform.

3 Tools

ENS3D development uses a Microsoft Visual Studio.NET 2003 environment. The choice for this tool is due to its compatibility with the libraries used in the system implementation. The libraries chosen for the implementation are open source ones, reducing the costs of development significantly.

The development has required the integration of a reasonable amount of libraries. However, the use of open source components has allowed an easier system customization. Although the system is being developed to use an integrated spatial database, it can be easily adapted to data reading in different archive formats.

Below one can find the main tools used for the implementation of the system's modules.

- **Spatial database:** ENS3D is expected to use a great amount of database manipulation, which includes the geographical information and the energy distribution networks of an entire city. Because of this a free licensed SDMS (*Spatial Database Management System*) was chosen, in this case PostgreSQL/PostGIS.
- **2D environment:** For the implementation of this environment it was chosen an OGR Library (group of libraries based on C++ codes which allow the reading and manipulation of different types of vector data) for spatial database reading. GDI+ (*Graphics Device Interface Plus*), more recent version of GDI, one of the main MS-Windows sub-systems, was used for the creation of the spatial database geometry.
- **3D environment:** In order to create the 3D GIS environment, it was chosen a group of VTP (*Virtual Terrain Project*) open source libraries which comprise two other libraries: vtlib and vtdata. Vtdata uses GDAL (*Geospatial Data Abstraction Library*) [5], OGR (GDAL sub-library for vector data manipulation) and PROJ.4 (for projection system configuration), among others. Vtlib uses OpenGL, libMini and OSG (*Open Scene Graph*) libraries for the rendering of vector data obtained by vtdata [10].
- **UD environment:** For this implementation it was developed an algorithm based on graphs which scans the energy distribution network structure and establish connections among the elements. Based on this information, the drawings of the unifilar diagram (UD) (with the symbols representing the power grid)

are made with the use of MFC (*Microsoft Foundation Class*) windows manager.

- **Interface:** The MFC windows manager was chosen for the implementation of the system interface.

Picture 1 shows the relation among the main libraries:

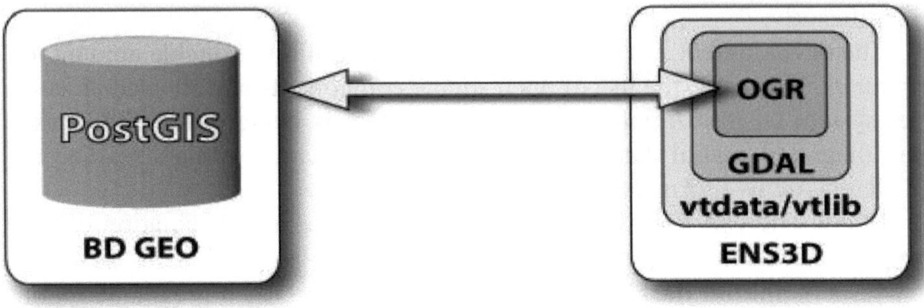

Pic. 1. Main libraries used in ENS3D.

4 ENS 3D Project

The basic elements for the system's development will be shown in this section.

4.1 Cartography

The digital cartographical map of São Carlos city was obtained after an agreement with São Carlos city hall. This cartographical map is compatible with a 1:2.000 map scale.

Since this map is not compatible with GIS, it was necessary the closing of polygons, the overlapping of nodes according to the conceptual model. This work aims at the creation of a digital cartographical map for a 2D environment.

For the 3D environment a DEM (Digital Elevation Model) was created from digitalized elevation contours and elevation points also in a 1:2.000 map scale.

Another product acquired for the project was a Quickbird digital image with 0,6m spatial resolution, obtained from the fusion of images of pancromatic and multispectrum bands.

This high-resolution image was applied to a texture over DEM to generate a photorealistic visualization of the model. Over this platform it will be added 3D models of some buildings, aiming at creating referential points to help the location and contextualization of the urban area and also to show the possibility of developing an entire city in 3D. For this model it was chosen governmental buildings, plants, hospitals and buildings offering essential services in the city.

The thematic data (substations, poles, transformers, feeders, and the cabling) were structured and stored in a relational spatial database.

4.2 Virtual Environment

The graphical interface developed for the virtual environment is based on the use of three environments, in a way that they are synchronizedly updated in the camera position and the location of the electrical elements.

The interface was projected to provide more freedom to the user in a way that he can have different visualizations of the energy distribution network at all times. The interface also provides a more complete interaction when compared to the 2D environments.

The energy distribution network is formed by several electrical elements. Its representation is usually made through a scheme where the system's components are represented by simple symbols, using a UD. The importance of the UD is to provide, in a concise way, the most significant data of the power system as well as its topology. The power distribution companies usually rely on the UD map as an essential document in the power restoration process.

The graphical interface proposed (where the three environments are integrated) allows the visualization of the energy distribution network from different perspectives: the UD environment, which shows the unifilar diagram of the energy distribution network; the 2D environment, that shows a 2D geo-referenced navigation map of the city; the 3D environment, in which the city can be visualized in a three-dimensional model under the primary energy distribution network and its electrical elements (transformers, switches, cabling, protection system, etc.).

Basically, this interface was created thanks to three complementary elements: the top fixed bar: with functionalities in all environments; the three-parted space with windows to the UD, 2D and 3D environments, which can be maximized/minimized according to the usage and; the dynamic lateral bar, which shows the specific data of the window and the selected element.

Since each window has different functionalities, it was created a dynamic lateral bar which changes features according to the environment selected. In other words, when the user selects one of the windows, the specific tools are automatically available in this bar.

The top fixed bar shows the most general functions, which apply to all windows and are always visible. The three environments are synchronized at all times and are complementary, providing the user with an easy navigation through the virtual environment, fast location in the urban area and direct reference from the UD elements to the 3D environment. Picture 2 shows the system's virtual environment.

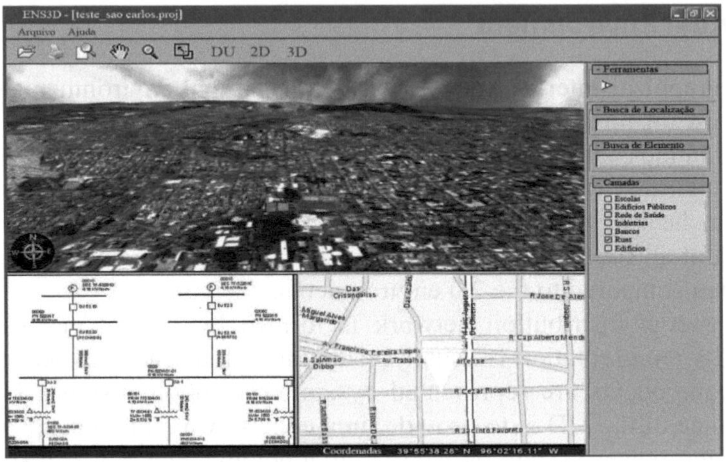

Pic. 2. ENS3D interface prototype.

4.3 Intelligent System

ENS3D system contains the restoration/reconfiguration module that executes the evolutive algorithm based on the GCR *(Graph Chain Representation)*. This algorithm collects the data related to the faulty sector and defines the best configuration based on this information, aiming at maximizing the amount of restored load. The algorithm exit report is graphically shown in the 3D environment, depicting the new condition of the switches after the operation. A final report with the results is given in the UD environment showing: the sequence of opening/closure of switches, the identification of the switches involved in the operation and their conditions. The user is also informed of the amount of energy consumed by the new configuration.

5 Implementation Details

In this section some of the phases involved in the systems implementation are described:

- Technical alternatives study: in this phase it was made a study on GIS Libraries for 2D and 3D graphical visualizations. The study revealed VTP as the most suitable group of libraries for the project in terms of learning curves, existing database compatibility and platform development and also due to its potentialities.
- Database generation: the database generation occurred in three phases:
    i. Use of script for the creation and population of tables that represent the system's assets.
    ii. Inclusion of the geometric column in the system's assets tables with the function PL of the PostGIS AddGeometryColunm. (In this phase, the modeled geometries have not been converted into the GEOMETRY format yet, allowing their storage in the records and their visualization).
    iii. Generation of thematic data table based on archives in the shape file format, using the shp2pgsql utility.
- Graphical visualization of thematic data: with the mfcSimple layout adaptation (VTP example project) and the thematic data table generation it was possible to access these tables with the use of OGR methods and generate the first graphical visualization of the layers of the system's interface prototype from the spatial database.
- Production of the preliminary digital terrain model with texture: a digital terrain model was created and applied as a texture to the satellite image of São Carlos city. Picture 3 shows the result of this process.
- Creation of 3D solids library: substations, power poles, transformers, etc, were modeled as solids with the use of 3D Studio Max. The pictures 4 and 5 show 3D models of electrical elements.

Pic. 3. DEM and QuickBird of São Carlos city.

Pic. 4. 3D models of power poles and transformers.

Pic. 5. 3D model of a substation.

6 Next Steps

ENS3D is an on-going project. Below are some of the functionalities to be implemented:

- Positioning of the energy distribution network elements in the 3D model of the city;
- Evolutive algorithm tests with real system data;
- Synchronization of the coordinates of the three environments of the system (2D, UD and 3D).

7 Conclusions

The integration of the intelligent system, the virtual environment and the database improves the visualization of the energy distribution network configuration, with its electrical elements and their location in a 3D environment. It is believed that this project contributes to a new concept of interface with a more interactive and intuitive approach.

It is important to mention the importance of the geoprocessing relying on a virtual environment as an essential tool to facilitate the decision making in real time and also the possibility of its integration in several areas in the power distribution companies. This concept could be extended to the application of this virtual environment for distance training, for instance.

The open source software community has been launching a series of products to cater for GIS users' needs. Despite the difficulties in integrating the developed open source tools, they are of great value in terms of cost reduction. In this sense, this work represents the possibility of developing a qualified GIS with the use of open source libraries.

It is important to note the application of some projects developed in 2D GIS in the power distribution systems [4] [7]. However, it is not known today any 3D GIS being effectively used, although there are initiatives applied to the power distribution systems [1].Therefore, this project aims at encouraging new research field initiatives.

Acknowledgments

This project had the financial support of FAPESP (*Fundação de Amparo à Pesquisa do Estado de São Paulo* – registration number: 2002/07862-3.

References

[1] Acero Marín, N.; Jurado Melguizo, F.; Rojas Sola, J. I. (2002). Nuevos métodos para la visualización del sistema eléctrico de potencia. In: XIV Congreso Internacional de Ingeniería Gráfica, Santander, Espanha, 5-7, jun. Disponível em: <http://146.83.6.25/software/FP/FPO3.pdf>. Acesso em: 30/05/2006.

[2} ANEEL. RESOLUÇÃO Nº 024, DE 27 DE JANEIRO DE 2000. Disponível em: <http://www.aneel.gov.br/cedoc/RES2000024.PDF>. Acesso em: 30/06/2006.

[3} ANEEL. RESOLUÇÃO Nº 223, DE 29 DE ABRIL DE 2003. Disponível em: <http://www.aneel.gov.br/cedoc/res2003223.pdf>. Acesso em: 30/06/2006.

[4] Crispino, F. Reconfiguração de Redes Primárias de Distribuição de Energia Elétrica utilizando Sistemas de Informações Geográficas. (2001). Dissertação de Mestrado. Escola Politécnica. Universidade de São Paulo.

[5] GDAL. GDAL Raster Formats. Disponível em: <http://www.gdal.org/formats_list.html>. Acesso em: 10/05/06.

[6] Gomes, F. V. Reconfiguração de Sistemas de Distribuição utilizando Técnicas de Otimização Contínua e Heurística para a Minimização de Custos. (2005). Tese de Doutorado. Universidade Federal do Rio de Janeiro.

[7] Pinto, A. M.; Machado, C. M. M.; Andreazza, C. J; Pereira, C. N.; Valério, E. C.; Correa, G. C.; Molina, J. R. C.; Nóbrega Filho, L.; Schelbauer, P. S.; Oliveira, R. P. (2004). Geoprocessamento para Uso no Campo em Apoio às Funções de Cadastro da Rede Elétrica e Fiscalização de Obras. In: SENDI 2004 - XVI Seminário Nacional de Energia Elétrica, Brasília-DF.

[8] Rahman, A. A. (2002). 3D GIS: the state-of-the art. In: International Symposium & Exhibition on Geoinformation 2002, Kuala Lumpur, Malásia. ISSN 1394 – 5505.

[9] Stoter, J.; Zlatanova, S. (2003), 3D GIS, where are we standing? In: ISPRS Joint Workshop on Spatial, Temporal and Multi-Dimensional Data Modelling and Analysis, Québec, October, 8 p. Disponível em: <http://www.gdmc.nl/publications/2003/3D_GIS.pdf>. Acesso em: 20/03/2006.

[10] VTP. Welcome to Virtual Terrain Project. Disponível em: <http://vterrain.org/>. Acesso em: 23/05/2006.

[11] Zhu, Q.; Li, F.; Zhang, Y. (2005). Unified Representation of three dimensional city models. In: ISPRS Workshop on Service of Spatial Data Infraestructure, XXXVI (4/W6), 14-16 out., Hagzhou, China. Disponível em:

<http://www.commission4.isprs.org/workshop_hangzhou/papers/237-242%20Qingzhu_A052.pdf>. Acesso em: 20/09/2006.

[12] Zlatanova, S.; Rahman, A. A.; Pilouk, M. (2002a) Trends in 3D GIS Development. In: Journal of Geospatial Engineering, Vol. 4, No. 2, Dezembro, pp.71-80. Disponível em: <http://www.lsgi.polyu.edu.hk/staff/ZL.Li/vol_4_2/01_zlatanova.pdf>. Acesso em: 23/06/2006.

[13] Zlatanova, S.; Rahman, A. A.; Shi, W. (2002b) Topology for 3D spatial objects. In: International Symposium and Exhibition on Geoinformation 2002, 22-24 Out, Kuala Lumpur, Malásia. Disponível em: <http://www.gdmc.nl/zlatanova/thesis/html/refer/ps/SZ_AR_WS_02.pdf>. Acesso em: 27/08/2006.